Philipp Kolbitsch

Chemical looping combustion for 100% carbon capture

Philipp Kolbitsch

Chemical looping combustion for 100% carbon capture

Design, operation and modeling of a 120kW pilot rig

Südwestdeutscher Verlag für Hochschulschriften

Impressum/Imprint (nur für Deutschland/ only for Germany)
Bibliografische Information der Deutschen Nationalbibliothek: Die Deutsche Nationalbibliothek verzeichnet diese Publikation in der Deutschen Nationalbibliografie; detaillierte bibliografische Daten sind im Internet über http://dnb.d-nb.de abrufbar.
Alle in diesem Buch genannten Marken und Produktnamen unterliegen warenzeichen-, marken- oder patentrechtlichem Schutz bzw. sind Warenzeichen oder eingetragene Warenzeichen der jeweiligen Inhaber. Die Wiedergabe von Marken, Produktnamen, Gebrauchsnamen, Handelsnamen, Warenbezeichnungen u.s.w. in diesem Werk berechtigt auch ohne besondere Kennzeichnung nicht zu der Annahme, dass solche Namen im Sinne der Warenzeichen- und Markenschutzgesetzgebung als frei zu betrachten wären und daher von jedermann benutzt werden dürften.

Verlag: Südwestdeutscher Verlag für Hochschulschriften Aktiengesellschaft & Co. KG
Dudweiler Landstr. 99, 66123 Saarbrücken, Deutschland
Telefon +49 681 37 20 271-1, Telefax +49 681 37 20 271-0, Email: info@svh-verlag.de
Zugl.: Vienna, University of Technology, Diss., 2009

Herstellung in Deutschland:
Schaltungsdienst Lange o.H.G., Berlin
Books on Demand GmbH, Norderstedt
Reha GmbH, Saarbrücken
Amazon Distribution GmbH, Leipzig
ISBN: 978-3-8381-0605-2

Imprint (only for USA, GB)
Bibliographic information published by the Deutsche Nationalbibliothek: The Deutsche Nationalbibliothek lists this publication in the Deutsche Nationalbibliografie; detailed bibliographic data are available in the Internet at http://dnb.d-nb.de.
Any brand names and product names mentioned in this book are subject to trademark, brand or patent protection and are trademarks or registered trademarks of their respective holders. The use of brand names, product names, common names, trade names, product descriptions etc. even without a particular marking in this works is in no way to be construed to mean that such names may be regarded as unrestricted in respect of trademark and brand protection legislation and could thus be used by anyone.

Publisher:
Südwestdeutscher Verlag für Hochschulschriften Aktiengesellschaft & Co. KG
Dudweiler Landstr. 99, 66123 Saarbrücken, Germany
Phone +49 681 37 20 271-1, Fax +49 681 37 20 271-0, Email: info@svh-verlag.de

Copyright © 2009 by the author and Südwestdeutscher Verlag für Hochschulschriften Aktiengesellschaft & Co. KG and licensors
All rights reserved. Saarbrücken 2009

Printed in the U.S.A.
Printed in the U.K. by (see last page)
ISBN: 978-3-8381-0605-2

TABLE OF CONTENTS

1	**INTRODUCTION**		**1**
	1.1	The greenhouse effect, carbon dioxide and climate change	1
	1.2	Greenhouse gas emissions and emission reduction	5
	1.3	Carbon capture and storage (CCS)	9
		1.3.1 Technologies for CCS	10
		1.3.2 CO_2 transport and storage	12
	1.4	Objective of this work	14

2	**CHEMICAL LOOPING COMBUSTION (CLC)**		**17**
	2.1	Principle of chemical looping combustion	17
	2.2	Oxygen carriers	19
	2.3	History of chemical looping combustion	21
	2.4	Current CLC experience with gaseous fuels	23
	2.5	Chemical looping reforming (CLR)	24
	2.6	Chemical looping combustion of solid fuels	26

3	**DESIGN OF CHEMICAL LOOPING COMBUSTORS**		**29**
	3.1	Reactor systems for chemical looping combustion	29
	3.2	Reactor systems with two interconnected fluidized beds	30
	3.3	The dual circulating fluidized bed (DCFB) reactor system	32
	3.4	Theoretical background on fluidization regimes in gas-solid fluidized beds	36
		3.4.1 The Geldart classification of particles	36
		3.4.2 Fluidization regimes in gas-solid fluidized beds	37
	3.5	Technical background for the design of chemical looping combustors . . .	42
	3.6	Design of a 120 kW CLC pilot rig	44
		3.6.1 Reactor system	45
		3.6.2 Reactor cooling system	48
		3.6.3 Auxiliary units	49

4	**OPERATING RESULTS OF THE CLC PILOT RIG**		**51**
	4.1	Hydrodynamic operation of the DCFB reactor system	51
		4.1.1 Pressure profiles of air and fuel reactors	51

    4.2 CLC performance with different oxygen carriers 57
            4.2.1 Experimental procedure and evaluation of results 57
            4.2.2 Performance of ilmenite for chemical looping combustion 59
            4.2.3 Performance of Ni-based particles 60
            4.2.4 Carbon formation in the fuel reactor 64
    4.3 Summary and outlook 65

5 MODELING OF THE CLC PILOT RIG   67

    5.1 Introduction 67
    5.2 Model development 68
            5.2.1 Model structure 68
            5.2.2 Reaction model 68
            5.2.3 Fluid dynamic model 70
            5.2.4 Energy balance 73
    5.3 Modeling results 73
    5.4 General aspects on modeling of continuous looping systems 75

6 CONCLUSIONS AND OUTLOOK   79

7 LIST OF PUBLICATIONS   81

8 NOTATION   83

    8.1 Abbreviations 83
    8.2 Symbols 84
    8.3 Sub and superscripts 85

9 REFERENCES   87

1

INTRODUCTION

1.1. The greenhouse effect, carbon dioxide and climate change

In recent years, the greenhouse effect has become well known from observations of climate change. Its principles and scientific background, however, have already been described in the early 19th century by the French mathematician and physicist Jean Baptiste Joseph Fourier. In his *Mémoires sur les températures du globe terrestre et des espaces planétaires* (1827) [1], he discussed the effect of the gaseous atmosphere on the temperature of the Earth's surface and mentioned *the effects of human industry* as impact on the climate. For the first time, the principle of trapping energy from visible light from the Sun as heat by different gases was introduced. Later in the 19th century, John Tyndall made experiments with different gases to clarify if this was really possible. In 1861, he published an article in which water vapor (H_2O) and carbon dioxide (CO_2) were identified as such gases [2]. By 1896, Svante Arrhenius put the effect of CO_2 in the atmosphere into numbers [3]. At this time it was known that some thousand years ago, major parts of Europe, North America and Asia were covered with ice. The reason for the cause of this ice age, however, was unknown. In his calculations he pointed out that the surface temperature of the Earth would increase by $8-9\,°C$ if the CO_2 concentration was increased by a factor of $2.5-3$. On the other hand, a decrease of the CO_2 concentration to $55-62\,\%$ of the value at that time would result in a temperature drop of $4-5\,°C$. At this time, this was no reason for concern. Arrhenius estimated that it would take approx. 3000 years of CO_2 production from industry to have an impact on the climate. The impact of greenhouse gases (GHG) in the atmosphere and the dramatic increase of emissions, however, were underestimated at that time.

Today, this opinion has undergone major revision. It is widely accepted that anthropogenic CO_2 emissions influence the current climate to a large extend or to put into the words of the International Panel on Climate Change (IPCC): *Most of the observed increase in global average temperature since the mid-20th century is very likely due to the observed increase in anthropogenic GHG concentrations* [4]. The concentration of greenhouse gases in the atmosphere has dramatically increased in the 20th century. The

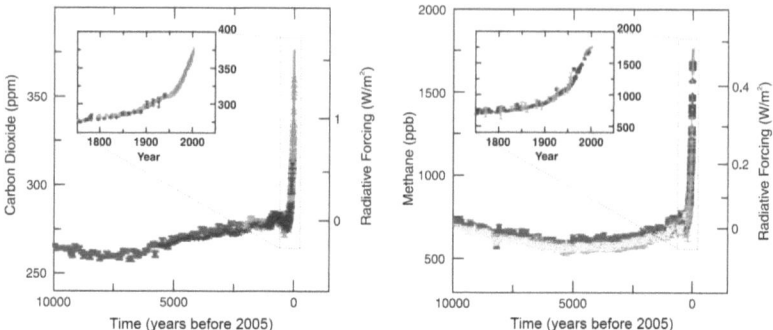

Figure 1.1: Atmospheric CO_2 and CH_4 concentrations in years before 2005 [4]

trend of atmospheric CO_2 and CH_4 concentrations in the atmosphere are illustrated in Figure 1.1 over a time-scale of more than 10000 years.

Compared to Arrhenius's model from 1896, many phenomena such as CO_2 uptake of the oceans and reduction of reflecting ice surfaces are taken into account in modern models. Nevertheless, it is impossible to prove that anthropogenic GHG emissions are the primary source for climate change. Organizations such as the International Panel on Climate Change (IPCC) or the United Nations Framework Convention on Climate Change (UNFCCC) collect data from different observations and refer to probabilities for possible causes of climate change. Beside this approach, one can simply take a look at different indicators of temperature changes in the past. The amount of deuterium present in antarctic ice probes, for example, has been found to be a marker for changes in local temperature. In Figure 1.2 the atmospheric concentrations of CO_2, CH_4 and N_2O are plotted with the deuterium variation in antarctic ice cores. This graph clearly identifies regions of warmer periods at high GHG concentrations. The consequences of our current situation for the global climate can only be estimated but could be catastrophic. Furthermore, the argument that the current GHG concentrations in the atmosphere could be caused by natural phenomena such as changes in the irradiation of the Sun alone is very much debilitated since the probability of only one such (unobserved) event withing 650,000 years is very low. Nevertheless, one has to keep in mind that this approach is far from being scientific and does not prove any connection between GHG and the current global warming. This might be one reason why some authorities still refer to natural reasons (volcano activity, change of irradiation of the Sun, etc.) for climate change and

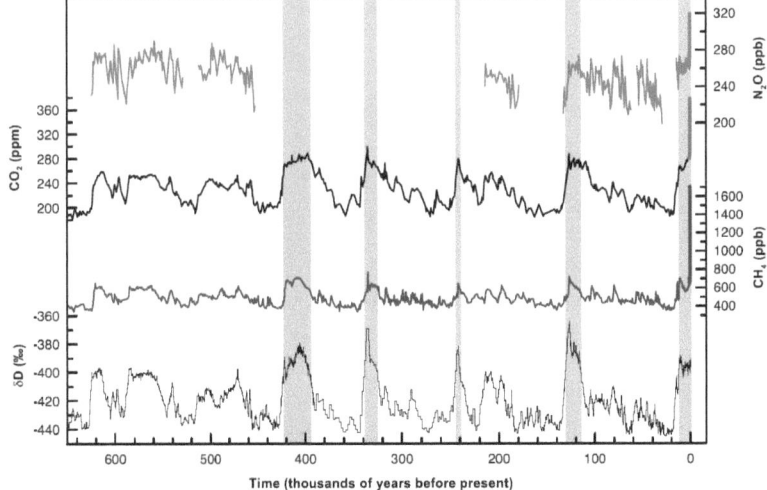

Figure 1.2: Atmospheric concentrations of CO_2, CH_4 and N_2O in air trapped withing ice cores and the variation of deuterium (δD), an indicator for local temperature, in antarctic ice over the last 650,000 years. Shaded bands indicate current and previous interglacial warm periods [4].

refuse to take proper measures to reduce GHG emissions.

Beside the debate on possible consequences of GHG emissions, observations of different markers of climate change are performed. In the last century, a steady increase of the average global surface temperature and the sea level as well as a decrease of snow cover is observed (see Figure 1.3). This temperature increase occurs all over the globe but is stronger at higher northern latitudes. It is even observed in the ocean at depths down to 3000 m [4].

These observations have led to serious concern and attempts to estimate the impact of increasing CO_2 emissions in the last decades and in the future have been initiated. On the basis of different scenarios for greenhouse gas emissions in the period from 2000 to 2100 (see IPCC SRES [5]) the further global temperature increase has been estimated. Depending on the SRES scenario applied, the models project a further temperature increase of $1.8 - 4.0\,°C$ and a sea level rise of $0.18 - 0.59\,m$ compared with the average values in the period 1980-1999 [4]. The IPCC further points out a number of expected

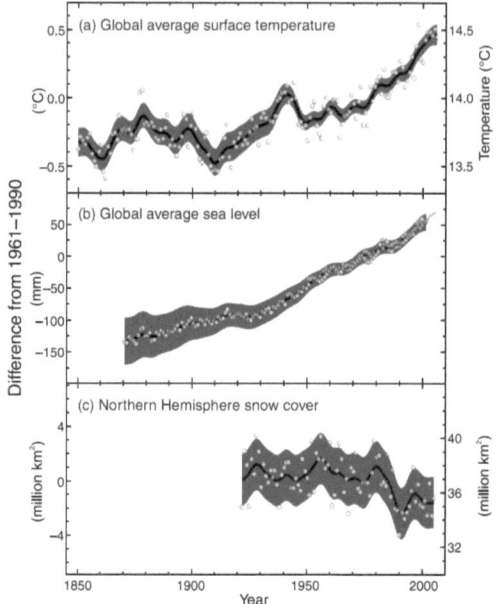

Figure 1.3: Observed changes of (a) the global average surface temperature, (b) the global average sea level and (c) the northern hemisphere snow cover for March-April. All reported differences are relative to average values in the period 1961-1990 [4].

climatic changes in the 21st century:

- The projected warming will be higher over land and at most high northern latitudes. Antarctica and the northern North Atlantic will have the least increases.

- Snow cover is expected to decrease and a thaw depth increase is projected for most permafrost regions. The amount of sea ice is projected to shrink in the Arctic and Antarctic.

- Extreme weather situations such as heat waves, tropical cyclones and heavy precipitation events will become more frequent.

- The amount of precipitation is projected to increase in high latitudes but decrease in most subtropical land regions.

The effects of GHG emissions are clearly visible in the current climate change. For the mitigation of these changes, the reduction of GHG emissions is indispensable. Possible options for the reduction of GHG emissions are discussed in the following sections.

1.2. Greenhouse gas emissions and emission reduction

Beside CO_2, other gases such as methane (CH_4) and nitrous oxide (N_2O) have been identified as greenhouse gases. Even though the concentration of these gases in the atmosphere is much smaller, their impact has to be considered owing to their high global warming potential (GWP) compared with CO_2 (see Table 1.1). To compare the effect of different GHGs, the CO_2 equivalent emission has been introduced. This parameter describes the amount of CO_2 that has the same global warming potential as a given mixture and amount of greenhouse gases, when measured over a specified time scale (100 years).[1]

Figure 1.4 shows the trend of GHG emissions since 1970. The total emissions have increased steadily. It is clearly visible that CO_2 from fossil fuel use has the biggest share on the overall emissions. In terms of sectors, energy supply and industry are the main contributors. For the mitigation of climate change, the ongoing trend of emissions will have to be reversed. Modern way of living, however, is very energy intensive and changes thereof are politically unpopular. Furthermore, many developing countries experience a significant economic upturn which is coupled with a strong increase in energy consumption and emissions (see Figure 1.5). This will be a very difficult challenge in which strong cooperations between industrial and developing countries will have to be established. As shown in figure 1.5, however, the OECD countries will also have a major share on energy consumption in the future.

Figure 1.6 shows the primary energy demand in terms of fuel type reported in the

industrial designation	chemical formula	lifetime [years]	GWP for given time scale		
			20 years	100 years	500 years
carbon dioxide	CO_2	see [4, 6]	1	1	1
methane	CH_4	12	72	25	7.6
nitrous oxide	N_2O	114	289	298	153
CFC-11	CCl_3F	45	6730	4750	1620
HFC-32	CH_2F_2	4.9	2330	675	205

Table 1.1: Global warming potential of some greenhouse gases [4]

---

[1]The GWP value depends on the decay of the gas concentration over time in the atmosphere. Therefore the global warming potential is meaningless without specified time scale.

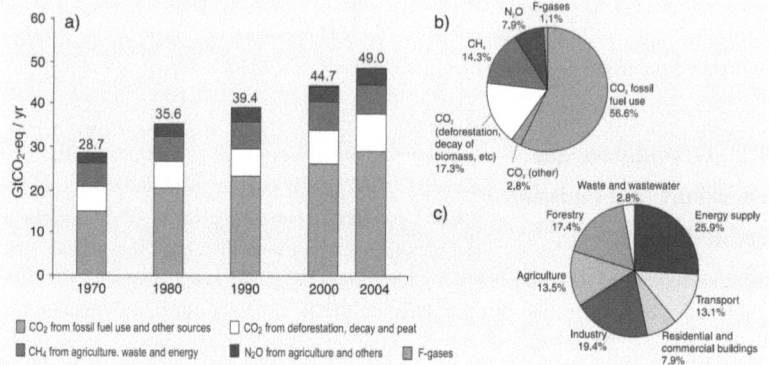

Figure 1.4: (a) Global annual emissions of anthropogenic GHGs from 1970 - 2004. Share of different anthropogenic GHGs (b) and sectors (c) in total emissions in 2004 in terms of CO_2 equivalent emissions [4].

World Energy Outlook [7] by the International Energy Agency (IEA) in 2006. It is clearly visible that fossil fuels had a big share in the past, have a big share in the present and will have a big share in the future. While the share of biomass, coal and hydro are projected to remain almost constant, a strong decrease of oil consumption is expected in

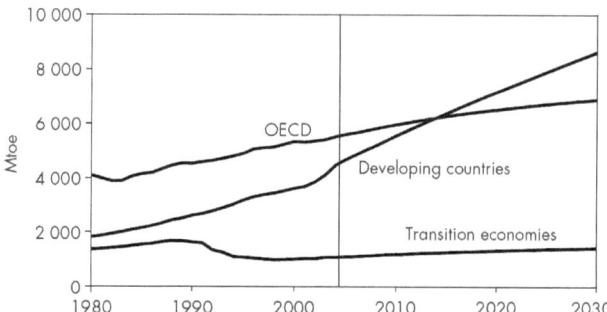

Figure 1.5: World primary energy demand by region since 1980 with an estimate until 2030 for the reference scenario. Over 70 % of the increase in world primary energy demand between 2004 and 2030 comes from developing countries [7].

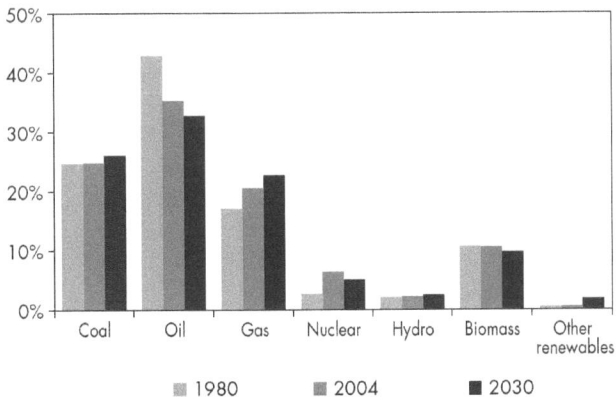

Figure 1.6: World primary energy demand by fuel in 1980, 2004 and an estimate for 2030 for the reference scenario. In 2004, approx. 80 % of the world primary energy demand was met by coal, oil and natural gas [7].

these scenarios. Also, a fuel switch toward natural gas is reported. Nuclear energy has experienced a strong increase in the past but is projected to decrease until 2030. Other renewables will experience a strong increase but the share on the total primary energy demand is still projected to be very low.

Nevertheless, there are many options for GHG emission reduction. Some are very simple, economic and state-of-the-art, others include major investments and additional research. The following list is just a brief overview of reduction possibilities:

- **Increase of conversion efficiency:**
  An efficiency increase in energy conversion allows equal electricity production with lower primary energy input. This measure also lowers fuel costs and is widely accomplished in industrial countries today. Recent advances in this sector include drying of lignite and the increase of live steam parameters (700 °C technology).

- **Replacement of fossil fuels by renewable fuels:**
  With this measure, fossil fuels such as coal are replaced with adequate renewables. Different kinds of biomass such as woody biomass, straw and other residues from food production and industry are discussed in this context. For a large scale fuel switch, energy crops will have to be grown which might get in conflict with food production.

- **Increased use of renewable technologies:**
  The use of renewable energy sources (beside biomass) has a huge potential but also has its limitations. Hydro power is already well developed and its further large scale potential is limited. Electricity production from wind is increasing very fast today and still has a large potential, especially for offshore wind farms. Other renewable energy sources such as tidal and wave energy, ocean currents and solar power have a great potential for the future but cannot yet compete with other technologies in terms of costs.

- **Fuel switch to less carbon intense fuels:**
  Depending on the amount of carbon in the fuel, the CO_2 emission per unit energy varies. Coal has a very high carbon content and therefore high CO_2 emissions. Natural gas (mainly lower C_xH_y), on the other hand, has a much lower carbon content. Additionally, natural gas can be converted with more efficient technologies such as combined cycle power plants. Compared to coal, however, natural gas is more expensive per unit energy and the resources are limited.

- **Nuclear energy:**
  Just like renewable fuels, nuclear energy produces no CO_2 emissions during operation (not including erection and disassembly of the power plant as well as mining). Its use, however, is declined by different nations worldwide owing to the problem of long-term storage of highly radioactive waste, the potential for nuclear accidents and the possibility for weapons-grade plutonium production. Also, with current technology, the potential for nuclear energy is limited owing to the decreasing resources of uranium. With modern technologies such as breeder reactors and the utilization of thorium as fuel, however, the potential is very large.

- **More effective energy utilization:**
  There is still a potential for energy saving. This includes an efficiency increase of many electronic devices and the minimization of heat losses of buildings with adequate insulation as well as the optimization of co-generation of heat and power.

- **Carbon capture and storage:**
  Carbon capture and storage (CCS) allows the utilization of fossil fuels without the emission of CO_2. The produced CO_2 is captured and stored and therefore not released into the atmosphere. Section 1.3 deals with CCS in more detail.

1.3. Carbon capture and storage (CCS)

As already shown in Figure 1.6, fossil fuels will continue to be a major player as primary energy source in the future. CCS allows the utilization of fossil fuels without emitting the produced CO_2 into the atmosphere. It could be applied in most CO_2 emitting sectors such as energy supply and industry. Even though CCS decreases the amount of CO_2 emitted into the atmosphere, the fuel input of the process is increased (see Figure 1.7).[2] Since the energy penalty for CCS is directly linked to the amount of produced CO_2 (and thus fuel input), highest possible efficiency of the process is essential to minimize the energy penalty for CCS. According to the IPCC [8], the fuel consumption increases by $11-22\%$ for natural gas combined cycle plants, $24-40\%$ for pulverized coal plants and $14-25\%$ for integrated gasification combined cycle plants (IGCC).

Many reports on carbon capture and storage have been prepared in recent years. The following introduction to CCS focuses mainly on the IPCC report on CCS [8], the IEA report on CCS [9] and the roadmap to carbon sequestration technology by the U.S. Department of Energy [10].

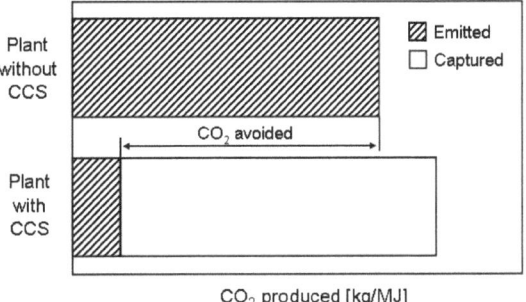

Figure 1.7: CCS decreases the amount of CO_2 emitted to the atmosphere but increases the fuel consumption and thus conversion efficiency and CO_2 production (adapted from [8]).

---

[2]The high energy demand for separation, compression and transport of the CO_2 eventually has to be compensated with additional primary energy.

1.3.1 Technologies for CCS

CCS for small scale applications, such as the transportation and building sector, is expected to be more difficult and expensive than from large CO_2 sources. Therefore, the large scale energy supply sector and industry have the biggest potential for cost effective CCS. Figure 1.8 gives a brief overview of the currently discussed CO_2 capture technologies.

In many industrial processes, CCS is already accomplished today. In the production of pure H_2, ammonia, alcohols and synthetic liquid fuels, CO_2 has to be removed from process gas streams. Most of this CO_2, however, is still vented today because there is no commitment for storage. In these processes, CCS could be integrated rather easily and cost effectively. One major advantage of these applications is the high partial pressure of CO_2 in the flue gase which makes the capture less energy intensive. Steel and cement production are further candidates for CCS from industrial processes [8].

Post-combustion capture applies conventional power boilers with an additional unit

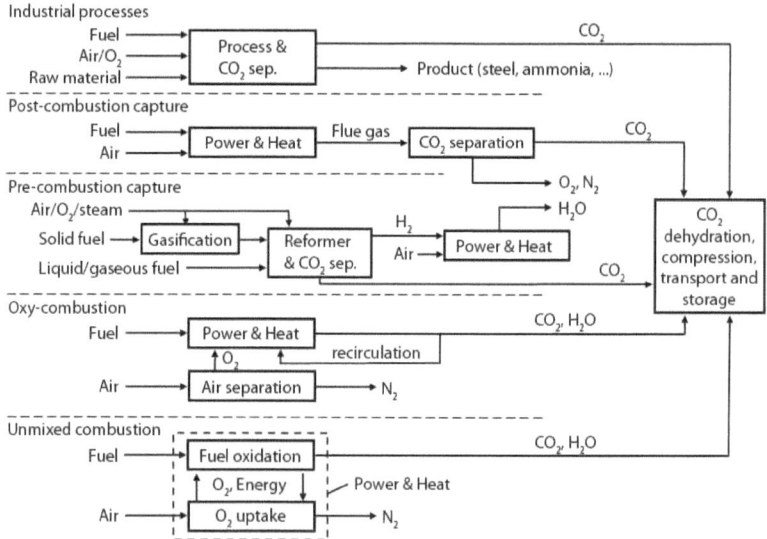

Figure 1.8: Discussed routes for CCS

for CO_2 removal from the flue gase. Owing to the low CO_2 concentration in the flue gas, amine scrubbers are applied. This process is approved and existing power plants could be retrofitted. The CO_2 separation unit, however, is very big owing to the high amount of flue gas to be processed. Further, the separation step has a high energy consumption. Post-combustion capture is primarily applicable to conventional coal-fired power plants but may also be applied to gas-fired units [10].

Pre-combustion capture has already been done for many years in the production of synthesis gas. The fuel is processed to syngas (CO + H_2) via gasification of solid fuels or reforming of gaseous or liquid fuels (steam reforming or partial oxidation) with subsequent water-gas shift reaction. This way, nearly all of the carbon in the fuel is processed to CO_2. Owing to the much higher CO_2 partial pressure in the syngas (high CO_2 concentration and absolute pressure) compared with flue gases of power plants, the gas separation unit decreases in size and cost. Furthermore, other processes such as the rectisol or selexol process, can be applied. Unfortunately, the fuel processing step is usually a high temperature process which implies a partial conversion of the chemical energy in the fuel to heat and therefore a loss of efficiency. Additionally, an air separation unit might be necessary for the supply of O_2 for gasification. This unit, however, would be much smaller than the air separation unit for oxy-combustion.

Oxy-combustion features the combustion of fuels with pure O_2 and recirculated flue gas. The primary products of such boilers are CO_2 and H_2O which can easily be separated. In this case, however, an air separation unit is necessary and additional research on the combustion with "artificial air" is necessary. Alternatively, less energy intensive air separation technologies are currently under research but are still not reliable today. According to the U.S. Department of Energy [10], oxy-combustion offers several additional benefits. Besides a $60-70\,\%$ reduction of NO_x emissions (very low N_2 content in combustion chamber) and increased mercury removal, retrofitting of existing coal-fired power plants is possible.

Unmixed combustion represents a fourth possibility for CCS. In this case, air and fuel are not mixed during combustion and two separate flue gases are obtained [11]. While the first stream usually contains the depleted air (N_2 with some excess O_2), the second stream contains the oxidized fuel (ideally CO_2 and H_2O). Solid oxide fuel cells (SOFC) and chemical looping combustion (CLC) are two candidates for this technology for CCS.

A comparison in terms of performance and cost of post-combustion capture, pre-combustion capture and oxy-combustion has been made by Davison [12].

1.3.2 CO_2 transport and storage

After the capture of CO_2, its transport to storage or utilization sites is necessary. This will be accomplished in high-pressure pipelines to onshore sites or by tankers (similar to existing LGP carriers) to offshore sites. The transport of CO_2 has been demonstrated in the U.S. since the early 1980s. Today, about 3000 km of land-based CO_2 pipelines are in operation around the world but primarily in North America [9].

For storage of CO_2, ocean storage and underground geological storage have been proposed. The latter can be divided into storage in deep saline formations, depleted oil and gas reservoirs and deep, unminable coal seams. The estimated capacity of these options is listed in Table 1.2.

For the geological storage, there exist three principle mechanisms for CO_2 sequestration. CO_2 can be trapped as gas or supercritical fluid under a layer of low-permeability rock, very similar to how natural gas is trapped in gas reservoirs. This mechanism is referred to as hydrodynamic trapping. The second mechanism is called solubility trapping in which CO_2 is dissolved into a fluid. CO_2 can also react with minerals and organic matter to become part of the solid mineral matrix (mineral trapping) [9].

Underground geological storage has been successfully put into operation in the Sleipner Project, operated by Statoil, in the North Sea (Figure 1.9). CO_2 from Sleipner West Gas Field is separated and injected into a large, saline formation 800 m below the seabed of the North Sea. Approximately 1 MtCO_2 is injected every year. The total storage capacity is estimated to be in the range of $1-10$ GtCO_2. A second example for geological storage is the In Salah - CO_2 Storage Project in Algeria where CO_2 is stored in a gas reservoir [8].

Enhanced oil recovery (EOR) has been utilized for many years, especially in the U.S. and Canada. For EOR, CO_2 is injected into oil reservoirs (Figure 1.10). This way, the

sequestration type	worldwide capacity [GtCO_2]
ocean	2300 (350) $-$ 10700 (1000)
deep saline formations	> 1000
depleted oil and gas reservoirs	675 $-$ 900
coal seams	3 $-$ 200
annual CO_2 emissions	26.4 ± 1.1 [4]

Table 1.2: Worldwide capacity of potential CO_2 storage sites compared with annual emissions (average value in the period 2000 to 2005). The potential for ocean storage is large relative to fossil fuel resources and is dependent on the atmospheric CO_2 stabilization concentration (number in brackets in ppm) [8].

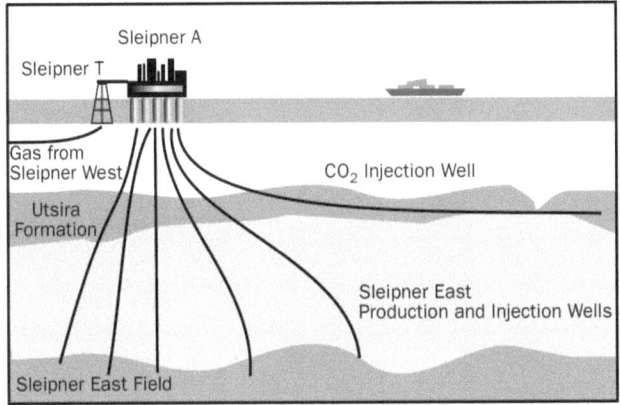

Figure 1.9: Simplified diagram of the Sleipner CO_2 Storage Project [9]

mobility of the oil and thus the productivity of the reservoir is increased. Applying EOR could therefore significantly reduce the price of CO_2 sequestration. An example for this technology is the Weyburn CO_2-EOR Project which is located in the Williston Basin, a geological formation in parts of the U.S. and Canada. It is expected that about $20\,\text{MtCO}_2$ will be stored in the field over the lifetime of the project ($20-25$ years). The dehydrated CO_2 is compressed and piped to Weyburn (southern Saskatchewan, Canada) for use in the field [8].

Current EOR activities mainly use CO_2 from natural resources. This involves additional emissions of CO_2 from these sources. A switch to CO_2 from carbon capture facilities is essential in the near future and will contribute to the reduction of current GHG emissions.

Oceans are part of the natural CO_2 cycle and already store parts of the anthropogenic CO_2 emissions. $500\,\text{GtCO}_2$ out of $1300\,\text{GtCO}_2$ total anthropogenic CO_2 emissions have already been taken up by the oceans in the past 200 years. This has resulted in a decrease of approx. 0.1 in pH value at the ocean surface but virtually no change at greater depths [8]. CO_2 can also be stored in form of lakes in depths where its density is higher than sea water (below $3000\,\text{m}$). The environmental impact of CO_2 lakes on aquatic life in the water and on the seabed, however, might be severe and still has to be evaluated.

Mineralization has also been mentioned in combination with CO_2 sequestration. The

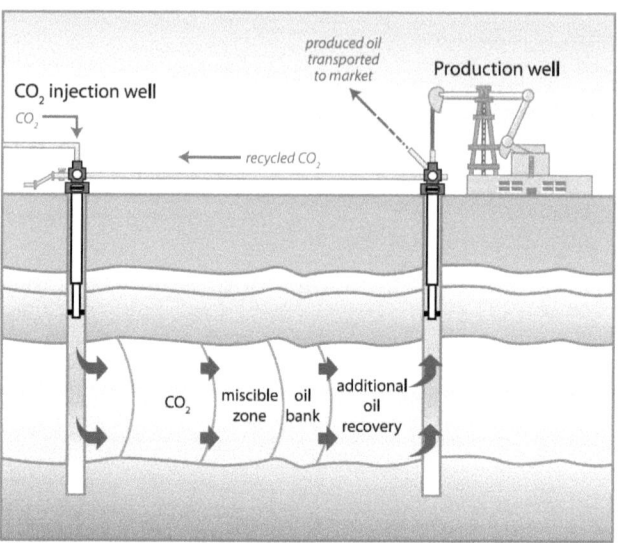

Figure 1.10: Injection of CO_2 for enhanced oil recovery. The CO_2 that is produced with the oil is separated and reinjected [8].

reaction of CO_2 with minerals (such as magnesium silicate) is energy producing and therefore energetically favorable. The formed carbonates are stable and environmentally harmless. This option, however, faces the challenge of huge amounts of materials that have to be handled and poor reactivity of the minerals. Further, this option of CO_2 sequestration implicates extensive mining activity which will have some environmental impact.

1.4. Objective of this work Chemical looping combustion is a very attractive technology for efficient energy utilization from fossil fuels with subsequent carbon capture. Up to now, CLC has been demonstrated on a scale up to 50 kW thermal power. This work deals with the design and operation of a 120 kW chemical looping combustor which features a new design concept for two interconnected fluidized bed reactors.

In the first part of this thesis the dual circulating fluidized bed (DCFB) reactor system is introduced. The DCFB reactor system offers a wide range of advantages compared with other systems. These include excellent gas-solids contact over the total reactor height, minimized solids inventory and high potential for scale-up. On the basis of this concept, the design of the 120 kW CLC pilot rig is made. This CLC unit can be fueled with natural gas and designed mixtures of H_2, CO, N_2, CO_2 and C_3H_8 from gas cylinders. The results presented in this first part are based on Paper I and Paper II.

The CLC pilot rig has been put into operation in early 2008. Since then, more than 100 h of hot CLC operation have been performed. In the second part of this work, different results from these experiments are presented and discussed. This part is based on Paper III, Paper IV and Paper V.

Beside the design and operation of the CLC pilot rig, a simulation tool for detailed modeling of reactor kinetics has been developed. The model is capable of characterizing gas and solids composition at different positions of the reactor system. Different operating cases are simulated and predictions for the pilot rig operation are made. The content of this part in based on Paper VI.

2

CHEMICAL LOOPING COMBUSTION (CLC)

2.1. Principle of chemical looping combustion

Chemical looping combustion is a novel two-step combustion process in which the mixing of fuel and air is completely avoided. The reactor system consists of two separate reactors, an air reactor (AR) and a fuel reactor (FR). The oxygen for combustion is transported by an oxygen carrier (OC) which circulates between the two reactors (Figure 2.1). Therefore, this process features inherent CO_2 separation and the energy penalty for the capture of CO_2 from the exhaust gas is minimized.

For the operation of a CLC reactor system, high gas-solids contact to assure appropriate fuel conversion (and minimize solids inventory) is obligatory. Further, transportability and continuous mixing of the particles is necessary. Therefore, both air and fuel reactors are typically designed as fluidized beds with the oxygen carrier as bed material.

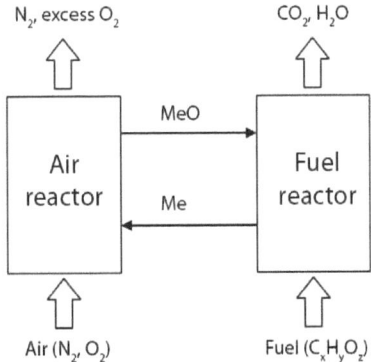

Figure 2.1: Principle of chemical looping combustion. Fuel and air are not mixed during combustion. The necessary O_2 is transported by an oxygen carrier from the air reactor to the fuel reactor.

Loop seals between the two reactors (typically fluidized with H_2O) avoid mixing of the reacting gases from each reactor. This design uses conventional technology that has been successfully implemented for many years in different applications such as solid fuel combustion and in different processes of petrochemistry.

The air reactor is usually designed as a fast fluidized bed to assure appropriate solids circulation. For the fuel reactor, bubbling and turbulent fluidized beds have been proposed. Bubbling fluidized beds offer a simple design but suffer from the possibility of fuel bypass in the bubble phase and limited fuel power of the plant. In turbulent fluidized beds, higher mechanical stress on the particles is expected (e.g. in cyclone separators). This approach, however, minimizes the fuel bypass and allows larger plant size.

In the fuel reactor, the fuel is oxidized according to

$$C_xH_yO_z + (2x + \frac{y}{2} - z)MeO_\alpha \rightleftharpoons xCO_2 + \frac{y}{2}H_2O + (2x + \frac{y}{2} - z)MeO_{\alpha-1} \quad (2.1)$$

reducing the oxygen carrier at the same time. The reduced oxygen carrier is then transported back to the air reactor where it is reoxidized according to

$$MeO_{\alpha-1} + \frac{1}{2}O_2 \rightleftharpoons MeO_\alpha. \quad (2.2)$$

The air reactor exhaust gas contains N_2 and some excess O_2, depending on the global air/fuel ratio. The reaction in the air reactor is always exothermic, whereas the reactions in the fuel reactor are slightly exothermic or endothermic depending on the active metal of the oxygen carrier and the fuel. The sum of reactions in air and fuel reactors is identical with the direct oxidation of the fuel, i.e.

$$C_xH_yO_z + (x + \frac{y}{4} - \frac{z}{2})O_2 \rightleftharpoons xCO_2 + \frac{y}{2}H_2O \quad (2.3)$$

When an adequate oxygen carrier is used and enough residence time is provided, the fuel reactor offgas ideally consists of CO_2 and H_2O only. After the condensation of H_2O, a highly concentrated CO_2 stream is obtained. Therefore, chemical looping combustion has been identified as a high potential technology for CCS. Beside this advantage, CLC has also been mentioned in context with low NO_x combustion. Owing to the low combustion temperature and the flameless combustion in the air reactor, low NO_x production is expected which has been demonstrated by Ishida and Jin [13] and Ryu et al. [14].

In principle, all kinds of gaseous fuels, such as natural gas or synthesis gas, can be used for CLC. Naturally, there is a great interest for the combustion of coal in chemical looping combustors. Two different approaches toward solid fuel utilization in CLC are discussed in section 2.6.

2.2. Oxygen carriers

A number of different metals and their oxides have been mentioned in literature for use as oxygen carrier in chemical looping combustors. The most important candidates are Cu, Fe, Ni, Co and Mn. The main demands for particles to be used as oxygen carriers are:

- Thermodynamic capability for sufficient fuel conversion.
- Sufficient strength against attrition for use in fluidized beds.
- Low tendency for agglomeration.
- High oxidation and reduction reaction rates.
- Sufficient oxygen transport capacity (i.e. the maximum mass fraction of O_2 that can be transported by a particle) to limit the required solids circulation.
- Low tendency for coke formation.
- High availability and low price of the raw materials.
- Environmental harmlessness.

None of the mentioned metals features all desired properties. Therefore, a trade-off has to be found. To increase the stability, the active metals are often combined with inert oxides such as Al_2O_3, TiO_2 or yttria-stabilized zirconia [15].

Figure 2.2 shows the O_2 transport capacity R_0 of different redox systems. In terms of reactivity toward hydrocarbons (especially CH_4), NiO-Ni is the favorable redox system. For the combustion of syngas all of the mentioned redox systems have high or adequate reactivity. The transition of Fe_2O_3 to Fe has a very high O_2 transport capacity but it is probably not feasible for different reasons. Beside thermodynamic limitation of fuel conversion between FeO and Fe, it is very difficult to operate a CLC reactor between four different oxidation states of the metal. Ni, Co and Cu also have very high oxygen transport capacities. Co and Ni are very problematic in terms of health and environmental impact. Cu also faces this problem but to a much lower degree. In terms of availability and raw material price, iron oxides are certainly superior to the other materials. The melting temperature of most redox systems is sufficiently high for the operation of CLC

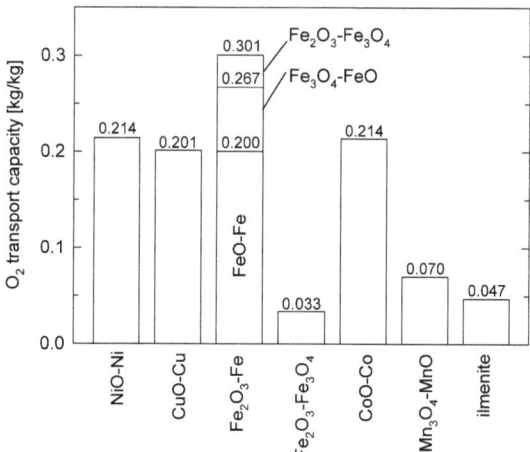

Figure 2.2: O_2 transport capacity R_0 of different unsupported materials discussed as oxygen carriers. Depending on the amount of added support material, R_0 is significantly reduced. The composition (and thus R_0) of ilmenite is dependent on the mining site.

combustors. Only Cu might face some problems with particle agglomeration owing to its low melting temperature ($T_{melt,Cu} \approx 1084\,°C$, $T_{melt,CuO} \approx 1201\,°C$).

Natural minerals, such as ilmenite, usually have a low oxygen transport capacity owing to the high content of inert substances (in this case TiO_2). Further, the BET surface is very low (very low porosity) which significantly reduces the reactivity of these materials. Owing to the very low price and high availability, however, these materials are also very interesting for CLC, especially for the combustion of syngas.

The combination of different active metals in oxygen carriers, so-called mixed oxides, have been proposed to combine different properties of used metals. In this context, Cu-Ni [16], Co-Ni [17] and Ni-Fe [18] based oxygen carriers have been mentioned. The combination of natural minerals with fabricated particles also belongs to this category. The reactivity of natural minerals toward H_2 and CO is usually quite adequate but for the conversion of hydrocarbons (especially CH_4) catalytic active particles are required. Therefore, the combination of these cheap natural minerals with a small fraction of highly reactive (and catalyzing) particles (e.g. Ni-based) seems very promising. The active Ni in the carrier is capable of reforming hydrocarbons to a syngas which is subsequently fully oxidized by the natural mineral.

There are numerous publications on the production and performance of different oxygen carriers. A theoretical analysis of 27 different oxide systems for the applicability in chemical looping combustion is presented by Jerndal et al. [19]. The investigation of 240 samples composed of Cu, Fe, Mn or Ni oxides on different support materials is made by Adanez et al. [20]. There are numerous further publications on the performance of different oxygen carriers. Johansson et al. [21] have prepared a comprehensive overview of most of these investigations.

2.3. History of chemical looping combustion

The principle of CLC was first mentioned in a patent by Lewis and Gilliland [22] as early as in 1954. In this patent, CLC with a Cu-based oxygen carrier was applied for the production of pure CO_2 from solid, oxidizable carbonaceous material (e.g. coal). This method had advantages over CO_2 recovery from flue gases from combustion and alcohol fermentation. The process applied a multi-stage bubbling fluidized bed reactor with counter flow of the solid fuel and the oxygen carrier. The spent oxygen carrier was then reoxidized in a regenerator. Lift gases were used to transport the solids between the reactors. The process was never commercialized but the patent already offers many details on oxygen carrier configuration, fuel handling, operating parameters (temperature, fluidization, ...), gas purities and reactor design (see Figure 2.3).

Later, in 1968 and 1983 the principle of chemical looping combustion was discussed by Knoche and Richter [23] and Richter and Knoche [24] with the intention to increase the reversibility of combustion processes and thus the thermal efficiency of the process. When applying adequate oxygen carries, CLC features a so-called chemical heat pump, i.e. the shifting of low temperature heat to a higher level. This is achieved with an endothermic fuel reactor and a highly exothermic air reactor. With this measure, the entropy gain of the overall process is reduced. In 1987, Ishida et al. [25] applied this theoretical concept to a power process including the evaluation of the thermal efficiency by means of graphic exergy analysis.

The link between CLC and CO_2 capture was first made by Ishida and Jin [26] in 1994. In this work, CLC is proposed for the improvement of the overall system efficiency of a power cycle and the recovery of H_2O and CO_2. This concept was carried on by Lyngfelt et al. [15] who proposed the first reactor concept for chemical looping combustion (Figure 2.4). In this study, the air reactor is operated in the fast fluidization regime to ensure sufficient particle transport to the fuel reactor, which is operated in bubbling regime.

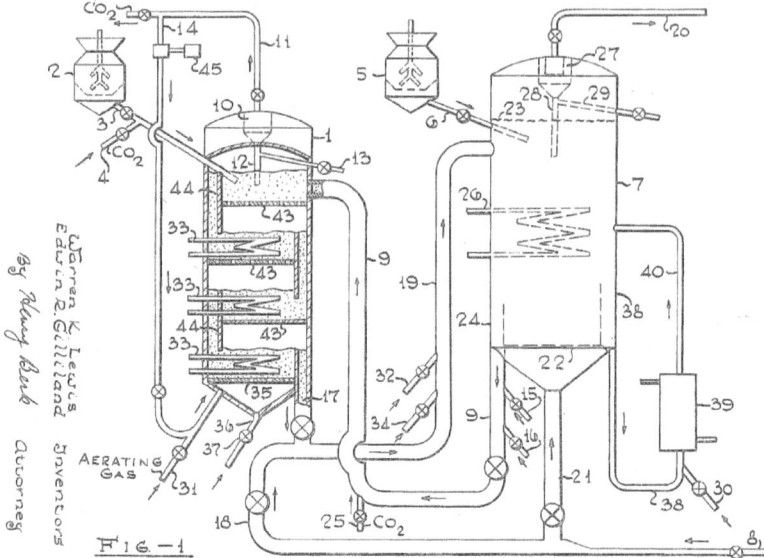

Figure 2.3: Process setup for the production of pure CO_2 using the principle of CLC. Cu-based particles were applied as oxygen carrier in this process. A detailed description and a list of all components is presented by Lewis and Gilliland [22].

Two loop seals, fluidized with H_2O, in the upper connection and the lower connection respectively, avoid mixing of the air and fuel reactor offgases. One major advantage of this concept is, in principle, the possibility of retrofitting existing circulating fluidized bed (CFB) boilers to CLC boilers. The combustion zone of the CFB boilers could be used as air reactor whereas the fuel reactor could be placed in the return leg. Some structural alterations in the offgas coolers, however, will certainly be necessary to maintain proper heat integration in the process.

The design by Lyngfelt et al. [15] is made for an atmospheric boiler with 10 MW fuel power and CH_4 as fuel. All major parameters for the design of a CLC unit are discussed and determined. Furthermore, important relations for the design and operation of this process regarding oxygen carrier performance, temperatures and reaction rates are outlined and discussed.

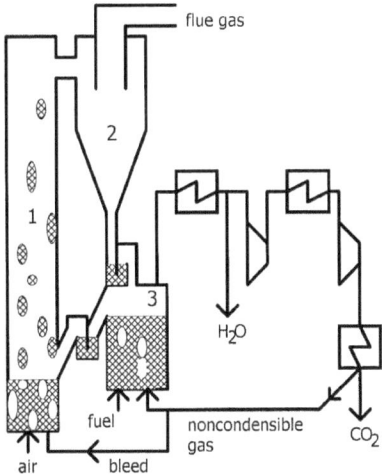

Figure 2.4: CLC reactor concept proposed by Lyngfelt et al. [15]

2.4. Current CLC experience with gaseous fuels

Chemical looping combustion for gaseous fuels has been demonstrated with various oxygen carriers in recent years. First experiments have been performed with batch reactors and thermo-gravimetric analyzers (TGA) to determine the reactivity of different oxygen carriers and their potential for use in CLC processes. Experiments in continuous looping units, however, are essential to prove the applicability of the CLC technology. In this sense, different continuous looping units have been erected and operated in recent years. An overview of some currently existing continuous looping units is shown in Table 2.1. In total, more than 1500 h of continuous looping operation have been reported worldwide in these test rigs. The units at Chalmers University of Technology and CSIC have reported more than 1000 h and 300 h respectively [27].

Batch reactors and TGAs were used in the beginning of CLC investigation and still are very useful tools today. Results of these experiments for gaseous fuels have been presented by Cho et al. [33, 34, 35], Corbella et al. [36], de Diego et al. [37, 38], Ishida et al. [39] and Johansson et al. [18, 40]. Continuous looping unit operation results with gaseous fuels have been published by Abad et al. [41, 42], Adanez et al. [16], de Diego

23

	location	power [kW$_{th}$]	fuel (ng ... natural gas)	reference (e.g.)
1	Chalmers	10	ng	[28]
2	Chalmers	10	RSA coal, pet coke	[29]
3	Chalmers	0.3	ng, syngas	[30]
4	CSIC	10	ng	[31]
5	South Korea	50	ng	[14]
6	South Korea	1	CH$_4$	[32]
7	Vienna (TUV)	120	ng, syngas, C$_3$H$_8$	Paper I

Table 2.1: Excerpt of the currently existing CLC units (data taken from Lyngfelt et al. [27])

et al. [31], Johansson et al. [43], Linderholm et al. [44] and Ryu et al. [14]. Reaction kinetics of different oxygen carriers have been determined by Abad et al. [45, 46], Garcia-Labiano et al. [47] and Zafar et al. [48, 49]. Garcia-Labiano et al. [50] have examined the effect of pressure on the performance of various oxygen carriers.

2.5. Chemical looping reforming (CLR)

The decrease of the global air/fuel ratio below 1.0 in a CLC process results in the emission of H_2 and CO from the fuel reactor. This is not desirable for chemical looping combustion but can be utilized for the production of a syngas (CO, H_2) from hydrocarbons. This process is known as chemical looping reforming and has been proposed by Mattisson and Lyngfelt [51] in 2001. It represents an alternative to partial oxidation and steam reforming of hydrocarbons and has been successfully demonstrated by, for example, Ryden et al. [52]. Production of H_2 with CLR in combination with a combined cycle process is an alternative to chemical looping combustion for fossil fuel energy utilization with CCS.

The straight-forward concept for CLR is the simple reduction of the global air/fuel ratio to the desired value, henceforth termed CLR(a) (Figure 2.5(a)). The global air/fuel ratio regulates the amount of produced H_2 and CO as well as the heat production of the process (see Figure 2.6). The operating region for autothermal reforming with respect to the global air/fuel ratio, is in the range of $0.25 - 0.40$, depending on heat losses and heat integration of the process. Below this value, the reactor system has to be externally heated to conserve the operating temperature.

CLR(a) offers a wide range of possibilities for process set-up and heat integration. An overview over different process set-ups and the evaluation of their performance and efficiency has been published by Ryden and Lyngfelt [53].

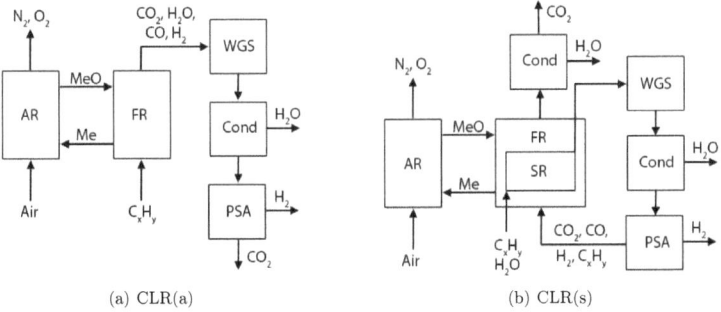

(a) CLR(a) (b) CLR(s)

Figure 2.5: Concept of CLR(a) and CLR(s) with condenser (Cond), pressure swing adsorption unit (PSA), steam reformer (SR) and water gas shift reactor (WGS).

A more advanced concept of CLR is the combination of steam reforming with CLC, henceforth termed CLR(s) (Figure 2.5(b)). In this concept, a conventional steam reformer is externally heated by a chemical looping combustor instead of direct heating by combustion of additional fuel. This process has the main advantage that the H_2 is produced at high pressure and pressure swing adsorption (PSA) for the removal of

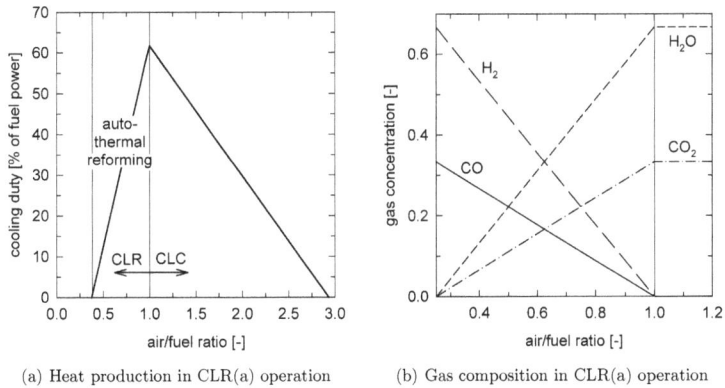

(a) Heat production in CLR(a) operation (b) Gas composition in CLR(a) operation

Figure 2.6: Heat production and product gas composition in CLR(a) operation (fuel: CH_4, $T_{FR} = 850\,°C$, $\triangle X_S = 0.10$, Ni based oxygen carrier with $40\,wt\%$ active NiO).

CO_2 is avoided. Nevertheless, a PSA for H_2 separation is necessary. Since some of the produced H_2 has to be combusted in the fuel reactor anyway, its separation efficiency can be somewhat lower. This concept is presented in more detail by Ryden and Lyngfelt [54].

2.6. Chemical looping combustion of solid fuels

Most investigations on chemical looping combustion have been performed on use of gaseous fuels. Coal, however, is cheaper and much more abundant than gaseous fossil fuels. Therefore, the reputation of CLC in the technology pool could increase very much if coal could be fired. In principle, there are two different approaches for use of solid fuels in CLC boilers:

1. **External gasification:** The combustion of syngas from coal gasification (mixture of CO and H_2) with CLC has been demonstrated by different researchers (e.g. [41, 42, 45], Paper III, Paper IV). Compared to hydrocarbons, this fuel is rather easy to convert owing to the high reactivity of many oxygen carriers toward CO and H_2. Therefore, the utilization of coal as fuel in chemical looping combustion with preceding gasification seems easily feasible. Unfortunately, the air separation unit (ASU) needed for coal gasification is very energy consuming and reduces the efficiency of the overall process. The amount of O_2 needed, however, is much smaller compared with for oxy-combustion. Additionally, this process is in competition with pre-combustion capture which is at a higher stage of development and implementation.

2. **In situ gasification:** The direct introduction of solid fuels into the fuel reactor would be very attractive in terms of simplicity and efficiency of the process. The solid fuel is in situ gasified in the fuel reactor in a H_2O/CO_2 atmosphere with subsequent oxidation of the produced CO and H_2. This approach, however, faces problems with ash separation from the particles in the fluidized beds and the transport of fuel particles together with the oxygen carrier to the air reactor where its combustion in an air atmosphere takes place. Since this transport will lead to CO_2 emissions from the air reactor and thus a decrease in the CO_2 capture efficiency, it has to be avoided. The reaction time for gasification is much higher than for combustion in the air reactor which implies that most of the fuel particles will eventually end up in the air reactor. In this case, the CO_2 is not captured and reduces the CO_2 capture efficiency. Therefore, either a unit (or mechanism) for

the separation of oxygen carrier and fuel particle in the connection between fuel and air reactors is needed or the design of the reactor system has to be changed. A staged fuel reactor design and a fuel reactor cascade are improved concepts for the in situ gasification of solid fuels in the fuel reactor. These improvements would significantly increase the residence time of the fuel particles in the reducing atmosphere and thus the CO_2 capture efficiency but also the complexity of the process. The direct introduction of solid fuels into the fuel reactor is a rather new discipline and has been studied by Berguerand and Lyngfelt [29, 55]. Much research is still necessary in this area.

3

DESIGN OF CHEMICAL LOOPING COMBUSTORS

3.1. Reactor systems for chemical looping combustion In recent years, different chemical looping combustors have been built all around the world (see Table 2.1 on page 24). The largest of these units with 50 kW fuel power is located in South Korea. In all of these units, fluidized bed reactor systems have been applied. Fluidized beds offer a wide range of benefiting aspects for use in chemical looping combustion. These include the possibility of solids transportation and excellent gas-solids contact and heat exchange properties.

For the design of a chemical looping combustor, different aspects have to be considered. High global solids circulation rate between air and fuel reactors is necessary to transport sufficient O_2 for the combustion. Further, high solids circulation minimizes the temperature difference between the two reactors and thus thermal stress. This is of special importance, when highly endothermic chemical reactions proceed in the fuel reactor (e.g. steam reforming reaction in chemical looping reforming). Further, the gas-solids contact in both reactors has to be maximized to ensure proper gas conversion. This is of special importance in the fuel reactor since the emission of unconverted fuel is hardly tolerable.

At Chalmers University of Technology, a CLC reactor system at a scale of 10 kW fuel power has been erected in 2002 and has meanwhile been operated for more than 1000 h [27]. The air reactor in this unit is designed as a fast fluidized bed. The entrained oxygen carrier particles are transported via a loop seal to a bubbling fluidized bed (BFB) fuel reactor (see Figure 3.1). The major priority in the design of this test rig was not to demonstrate the behavior of large CLC power plant performance in small scale but to obtain high operating flexibility. This was necessary to allow parameter variations to study particles and reactions in continuous operation and for the case that unforeseen changes had to be made [28].

BFBs suffer from the possibility of gas bypass in the bubble phase. In the particle-free freeboard of BFBs, no relevant reactions can be expected for CLC because the necessary oxygen carrier particles are missing. The gas slip can be minimized by low

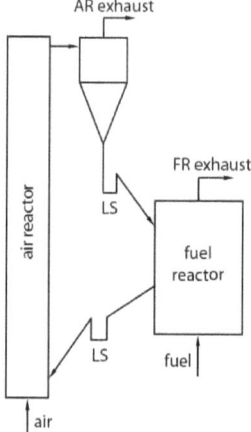

Figure 3.1: Set-up of the 10 kW CLC reactor system at Chalmers University of Technology. The air reactor is designed as a fast fluidized bed whereas in the fuel reactor a bubbling fluidized bed is applied (adapted from [28]).

fluidization numbers and sufficient bed height. This will result, however, in relatively large fuel reactor bed cross section areas, high solids inventors and high bed pressure drops. Circulating fluidized beds (CFBs) offer the presence of solids over the total reactor height. Therefore, gas-solid reactions are expected in the whole reactor volume. Compared with BFBs, increased gas conversion has generally been observed in CFBs [56]. Further, these reactors have a smaller cross section to power ratio which lowers the overall plant size compared with BFB boilers. Therefore, it seems promising to design the fuel reactor as fast or turbulent fluidized bed.

3.2. Reactor systems with two interconnected fluidized beds

Reactor systems with more than one fluidized bed have already been successfully commercialized. The classical application thereof is the fluid catalytic cracking process (FCC), in which one reactor is used for cracking higher hydrocarbons and the other reactor for the combustion of char formed on the catalyst particles (catalyst regeneration). Interconnected fluidized beds have further been applied for biomass gasification [57], carbonate loop-

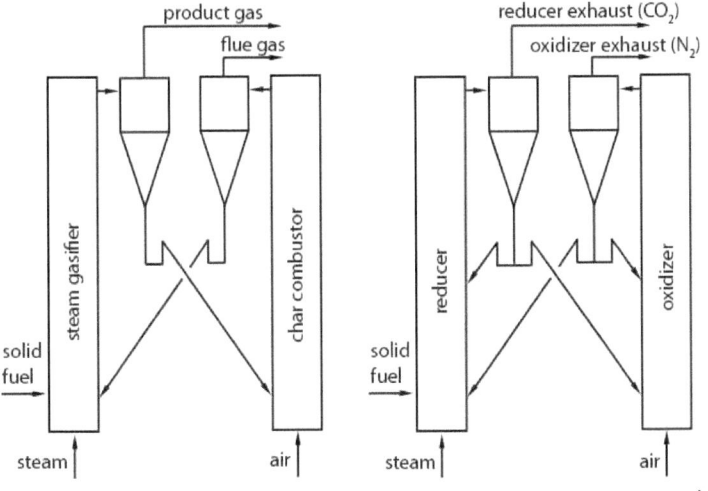

(a) Battelle/FERCO biomass gasifier [61]   (b) Alstom chemical looping reactor concept [62]

Figure 3.2: Reactor system concepts with two interconnected circulating fluidized beds

ing [58, 59] and sorption enhanced reforming [60].

Systems with two interconnected circulating fluidized bed reactors have been mentioned in literature by Paisley et al. [61] for biomass gasification (Figure 3.2(a)) and by Andrus [62] for chemical looping processes (Figure 3.2(b)). In these reactor system concepts, solids entrainment in both reactors is crucial in order to provide solids circulation between the reactors. Further, the solids have to pass two cyclone separators for one global solids loop. This will increase the fragmentation of the particles and the loss of fines which could be an economical disadvantage when expensive oxygen carriers are used. The solids hold-up in these reactor systems is defined by the fluidization regime in each reactor, which changes at different loads. This has to be considered in the design phase and proper measures have to be taken to avoid the accumulation of solids at different locations in the reactor system.

The dual circulating fluidized bed (DCFB) reactor system is an alternative to the concepts proposed by Paisley et al. [61] and Andrus [62]. In this reactor system concept, only one of the two reactors has an influence on the global solids circulation. Further, inherent solids inventory stabilization and reduced particle strain are expected.

3.3. The dual circulating fluidized bed (DCFB) reactor system

The DCFB reactor system is a novel way for interconnecting two circulating fluidized beds. The design has primarily been made to satisfy the demands for chemical looping combustion and chemical looping reforming but it can be applied in numerous applications where two circulating fluidized beds are combined. These processes include biomass gasification, sorption enhanced reforming and carbonate looping. The main requirements for a chemical looping reactor systems can be summarized as:

- **Excellent gas-solids contact:**
  For satisfactory gas conversion in the reactors, excellent gas-solids contact is necessary. This is of major importance in the fuel reactor where a significant slip of fuel is hardly tolerable. Since O_2 is provided in excess in the air reactor, the gas slip in this reactor is not as problematic as in the fuel reactor. An increased air/fuel ratio, however, has an impact on the overall efficiency of the process.

- **High global solids circulation:**
  The solids circulation rate controls the difference in degree of oxidation of the particles in each reactor. Garcia-Labiano et al. [47] have shown that the reactivity of the particles decreases with the residence time in each reactor.[1] A decrease of reactivity, however, will increase the minimum amount of solids required for the operation of the reactor. This implies that a maximized solids circulation rate will minimize the required amount of particles in each reactor. Further, the solids circulation rate influences the temperature difference between air and fuel reactors.

- **Low particle attrition:**
  Particle attrition has an influence on the operating cost of a CLC boiler. A minimization of particle attrition and fragmentation reduces the amount of particle renewal and thus costs, especially when expensive oxygen carriers are used.

The reactor system shown in Figure 3.3 satisfies these demands very well. In principle, this reactor system concept uncouples the impact of one of the reactors from global solids circulation. The solids are entrained in the air reactor, separated in a cyclone separator and directed to the fuel reactor via a loop seal (upper loop seal, ULS). A connection in the bottom part of both reactors, also executed as loop seal (lower loop seal, LLS), allows the back flow of the solids from the fuel reactor to the air reactor and closes

---
[1]This is associated with the distribution of oxidized and reduced sites in the particle and is described in more detail in section 5.4.

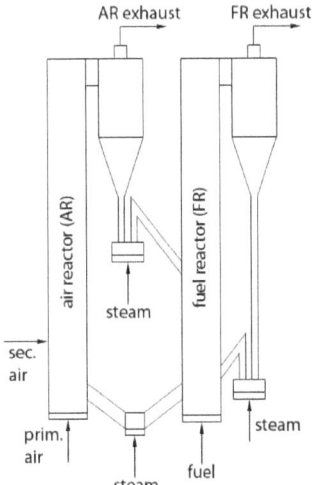

Figure 3.3: Sketch of the DCFB reactor system for chemical looping combustion. Both, air and fuel reactors are designed as CFBs with loop seals in between. The air reactor creates the driving force for global solids circulation. A connection in the bottom part of the two CFBs allows the solids to backflow to the air reactor and closes the global solids loop.

the global solids loop. The global solids circulation rate can effectively be controlled by staged fluidization in the air reactor which is common practice in CFB technology. The solids entrained from the fuel reactor are separated in a cyclone separator and directed back to a lower section of the fuel reactor via the internal loop seal (ILS). Therefore, the fuel reactor is uncoupled from the global solids loop and can be optimized in terms of fuel conversion or other aspects.

Most of the particle attrition and fragmentation in CFBs occurs in the cyclone separators. In the reactor systems mentioned by Paisley et al. [61] and Andrus [62], each particle has to pass at least two cyclone separators for one global solids loop. In the DCFB reactor system, every particle ideally has to pass only one cyclone separator for one global solids loop. Depending on the fuel reactor entrainment (fluidization number), the number of cyclone separators per global solid loop is slightly increased but will certainly be below 2.0 when the reactor system is properly designed. Thus, the DCFB reactor system features minimization of solids attrition and fragmentation compared

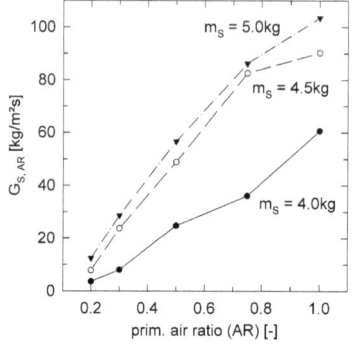
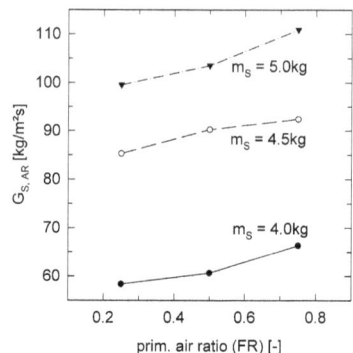

(a) Impact of air staging in the air reactor on $G_{S,AR}$

(b) Impact of air staging in the fuel reactor on $G_{S,AR}$

Figure 3.4: Effect of air staging in the (a) air reactor and (b) fuel reactor on the global solids circulation rate $G_{S,AR}$ in a cold flow model experiment. The operating conditions correspond to 120 kW fuel power and a global air/fuel ratio of 1.2 in the hot unit (see Pröll et al. [64] for a detailed description of the applied parameters).

with other interconnected CFB systems.

A further advantage of this reactor system is the inherent stabilization of the solids hold-up in the system. In this concept, the solids cannot accumulate at any point of the reactor system when proper fluidization of the CFBs and loop seals is provided. In fact, owing to the connection in the bottom part, a distinct stable solids level in both reactors is formed whose height is dependent on operating parameters and total solids inventory.

CFBs are currently built with fuel powers up to 600 MWel [63]. Since CFBs are the major components of the DCFB reactor system, very high scalability of this reactor system is expected.

A cold flow model has been designed to investigate the hydrodynamic behavior of the DCFB reactor system. The model represents the 120 kW CLC pilot rig at Vienna University of Technology with a linear geometric factor of 1 : 3. The design of the hot unit is described in section 3.6. Different parameter variations have been performed to determine the solids circulation rate at different positions of the reactor system. Figure 3.4 shows the impact of air staging in the air and fuel reactors on the global solids circulation rate. One can clearly see that the air reactor has the major impact on

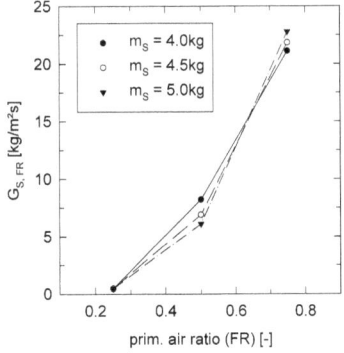
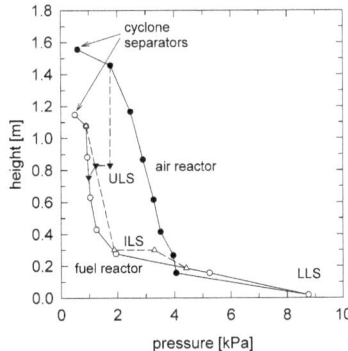

(a) Impact of air staging in the fuel reactor on $G_{S,FR}$

(b) Pressure profile of the DCFB reactor system.

Figure 3.5: Effect of air staging in the fuel reactor on the internal solids circulation rate $G_{S,FR}$ and pressure profile of the DCFB reactor system (cold flow model direct data). The operating conditions correspond to 120 kW fuel power and a global air/fuel ratio of 1.2 in the hot unit (see Pröll et al. [64] for a detailed description of the applied parameters).

the global solids circulation rate whereas the fuel reactor shows hardly any impact. This implies that the fuel reactor can be optimized with respect to fuel conversion without seriously affecting global solids circulation. On the other hand, the air reactor has to be designed in a way to assure solids transport at all times which might negatively affect gas conversion. Since full gas conversion is not necessary in the air reactor, however, this should be a minor problem.[2]

The effect of air staging in the fuel reactor is shown in Figure 3.5(a).[3] Just as in the air reactor, the solids circulation rate is strongly dependent on the fraction of air that is introduced in the lower section. On the one side, high solids circulation promotes the amount of solids in higher sections of the reactor and therefore promotes fuel conversion. On the other side, higher solids circulation in the fuel reactor increases particle attrition

---

[2] Combustors are usually operated with some excess of air. Therefore, some of the O_2 will be emitted from the air reactor anyway.

[3] In a hot unit this would correspond to fuel staging in the fuel reactor. Maximization of the gas residence time in the fuel reactor premises the introduction of all fuel in the lower section. Therefore, staging in the fuel reactor will probably not be considered in a commercial plant. It might be advantageous, however, to recirculate some fuel reactor exhaust gas to an upper section of the fuel reactor in some part load operating cases.

owing to the increased mechanical stress in the cyclone separator.

Pressure profiles of air and fuel reactors are shown in Figure 3.5(b). The very steep increase in the fuel reactor trend indicates a dense bottom region with a very steep increase of the void fraction in the upper section of the reactor. In the air reactor, however, a linear pressure profile is observed which indicates a constant void fraction over the total reactor height. Further, the three loop seals are sketched with their connections to the air and fuel reactors. It has to be mentioned that these profiles depend very much on the operating conditions of the single reactors and that this special case represents the CLC combustor at Vienna University of Technology. In principle, all sorts of different regimes can be applied in the air and fuel reactors as long as the global solids circulation is sufficient with respect to process requirements. The results of the cold flow model investigation have been summarized by Pröll et al. [64].

3.4. Theoretical background on fluidization regimes in gas-solid fluidized beds

In this section, the theoretical background on fluidization is briefly summarized. Beside a general classification of different particles, the major governing equations for the transitions between the different fluidization regimes are discussed. This extensive subject, however, is only touched to an extent that is absolutely necessary to distinguish the different fluidization regimes. For a more detailed introduction, the author refers to the comprehensive studies published by Grace and Bi [65] and Kunii and Levenspiel [66].

3.4.1 The Geldart classification of particles

Fluidization of particles is very much dependent on particle characteristics. Particle size and density influence the superficial velocity at which regime transitions occur to a high extent. Geldart [67] has introduced a classification in which the particles are distinguished in four different groups. From the smallest to largest, Kunii and Levenspiel [66] describe these groups as follows:

- Group C: cohesive or very fine powders which are extremely difficult to fluidize owing to great interparticle forces. Face powder, flour and starch are typical examples of group C particles.

- Group A: aeratable materials, or materials with a small mean particle size and/or low density. These particles fluidize easily with smooth fluidization at low gas velocities and controlled bubbling at higher gas velocities.

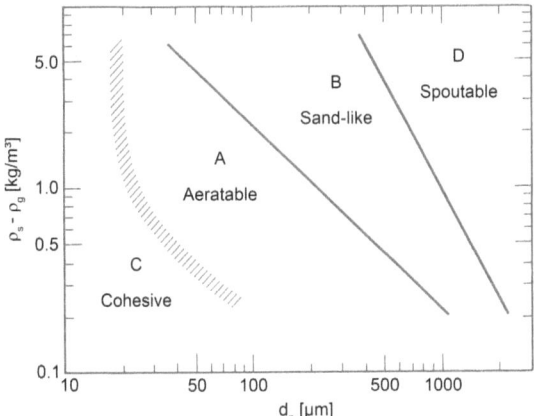

Figure 3.6: The Geldart classification of particles for air at ambient conditions (adapted from [66])

- Group B: sand-like particles with increased particle size and particle density ($40 < d_p < 500\,\mu$m and $1.4 < \rho_p < 4.0\,\text{kg/m}^3$, respectively). These particles fluidize well and show intense bubbling action with bubbles that grow large.

- Group D: spoutable or large and/or dense particles. Deep beds of these particles are hard to fluidize. Severe channeling and spouting behavior is observed when the gas distribution is very uneven. Many drying grains, coffee beans, gasifying coal and some roasting metal ores are examples of group D particles.

In Figure 3.6, the different Geldart particles are shown in a density vs. particle size diagram.

3.4.2 Fluidization regimes in gas-solid fluidized beds

When a gaseous fluid is introduced at the bottom of a fixed bed of particles, different hydrodynamic conditions will arise depending on the amount of introduced fluid. Starting from very little amounts of gas, the solids remain in a fixed position (fixed bed) until the minimum fluidization velocity is reached. From that point, the particles are fluidized. A further increase of the gas velocity expands the fluidized bed and causes, in most cases, the formation of bubbles (bubbling regime). These bubbles become larger with increased superficial velocity of the gas until the size of the bubbles becomes comparable with the column diameter at which point slugging occurs (slugging regime).[4] Once the force impressed on the particles by the gas stream outbalances its weight, the particles are elutriated. In order to keep a constant bed inventory, the elutriated particles have to be separated from the gas stream and then be recycled to the bed. When the standard deviation of the pressure fluctuations in the bed reaches a maximum, the turbulent regime emerges. Further increasing the gas velocity results in a continuous increase of the amount of elutriated particles. As soon as significant entrainment is observed, the regime is termed fast fluidization. In this regime there still exists an axial solids concentration profile with upward movement of the particles in the core and downward movement at the wall. At even higher gas velocities, pneumatic conveying emerges which is characterized by a loss of the axial variations of the solids concentration except in the bottom zone [68]. The mentioned flow patterns of gas and solids are illustrated in Figure 3.7. In the following, a set of equations is provided to distinguish between the described phenomena.

Fixed beds occur in the range $0 < U < U_{mf}$ with the minimum fluidization velocity U_{mf}. The gas pressure drop in a fixed bed is dependent on the gas velocity and can be quantified by the Ergun equation:

$$f_p = \frac{150}{Re_p} + 1.75 \qquad (3.1)$$

The friction factor f_p

$$f_p = \frac{\Delta p}{L} \frac{d_p}{\rho U^2} \left(\frac{\epsilon^3}{1-\epsilon}\right) \qquad (3.2)$$

and the particle Reynolds number Re_p

$$Re_p = \frac{d_p U \rho}{(1-\epsilon)\mu} \qquad (3.3)$$

depend on the pressure drop in the packed bed Δp, the bed length L, the void fraction of the bed ϵ, the equivalent spherical diameter of the particle d_p, the fluid properties (ρ and μ) and the superficial velocity U. Fluidized beds are characterized by a constant gas pressure drop which is determined by the gravitational force of the solids inventory in a column:

$$\Delta p = \frac{m \cdot g}{A} = (\rho_p - \rho_g)(1-\epsilon)gL \qquad (3.4)$$

From Equations (3.1) and (3.4), a relation for the minimum fluidization velocity, at

---

[4]This is only valid for sufficient bed height and Geldart A, B and D particles.

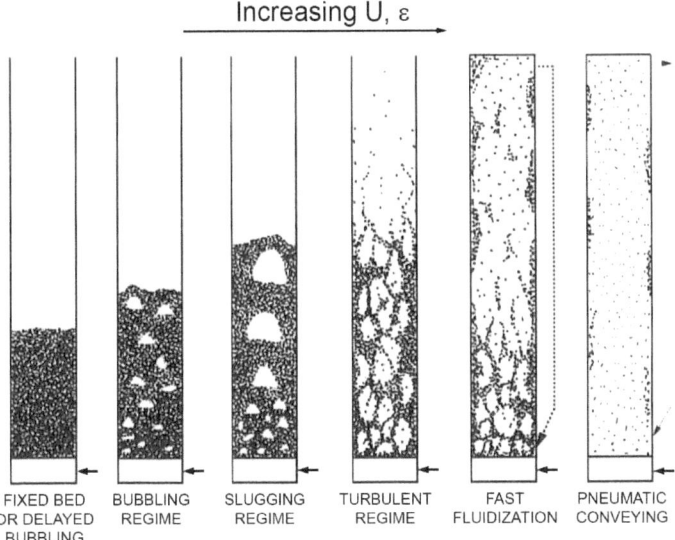

Figure 3.7: Gas-solids flow patterns in fluidized beds (adapted from [68])

which fluidization first occurs, can be derived in the form

$$Ar = c_1 \cdot Re_{mf} + c_2 \cdot Re_{mf}^2 \qquad (3.5)$$

with the Archimedes number

$$Ar = \frac{\rho_g \cdot d_p^3 \cdot (\rho_p - \rho_g) \cdot g}{\mu^2}. \qquad (3.6)$$

Equation (3.5) is a quadratic equation for the Reynolds number at minimum fluidization velocity and can be approximated as [65]

$$Re_{mf} = \sqrt{27.2^2 + 0.0408 Ar} - 27.2 \qquad (3.7)$$

with

$$Re = \frac{\rho \cdot d_p \cdot U}{\mu}. \qquad (3.8)$$

Starting from Re_{mf}, the bed is fluidized. The superficial velocity at which bubbling fluidization commences is highly dependent on particle properties. While Geldart C particle do not tend to bubbling fluidization at all, the minimum bubbling velocity U_{mb} is equal to U_{mf} for Group B and Group D particles. A region of bubble-free fluidization thus exists only for Group A particles and can be calculated as follows (note that it is essential to apply SI units in Equation (3.9)):

$$U_{mb} = 33 d_p \left(\frac{\rho_g}{\mu_g}\right)^{0.1} \tag{3.9}$$

For Geldart B and D particles, U_{mb} predicted from equation (3.9) is less than U_{mf}. Therefore, U_{mb} must be taken as equal to U_{mf} in this case [65]. When the superficial velocity is increased further, the bubbles become larger and eventually reach sizes comparable with the column diameter. At this point slugging occurs [65]. Stewart and Davidson [69] have estimated the minimum slugging velocity U_{ms} to

$$U_{ms} = U_{mf} + 0.07\sqrt{gD}. \tag{3.10}$$

According to Grace and Bi [65], however, slugging is not encountered in shallow beds (i.e. $H/D < 1$), in columns of very large diameters or for fine particles (i.e. $d_p < 60\,\mu\text{m}$) because bubbles are then unable to grow to sizes comparable to the column diameter.

The terminal velocity U_t, at which the single particles start to be elutriated, can be derived from balancing the particle weight, buoyancy and force due to friction and reads as

$$U_t = \sqrt{\frac{4}{3} \frac{\rho_p - \rho_g}{\rho_g} \frac{d_p \cdot g}{C_W}}. \tag{3.11}$$

The drag coefficient C_W of a particle is very much dependent on the Reynolds number. In the laminar region (Stokes region), C_W is calculated from

$$C_W = \frac{24}{Re} \quad (Re < 0.2), \tag{3.12}$$

in the turbulent region (Newton region) from

$$C_W = 0.43 \quad (Re > 1000) \tag{3.13}$$

and in the transition region, an implicit formulation for C_W is used:

$$C_W = \frac{24}{Re} + \frac{4}{\sqrt{Re}} + 0.4 \quad (0.2 < Re < 1000) \tag{3.14}$$

Formally, the turbulent regime initiates when the standard deviation of pressure fluctuations in the bed reaches a maximum ($U = U_c$). According to Bi and Grace [70], this transition first occurs when the relation

$$Re_c = 1.24 Ar^{0.45} \qquad (2 < Ar < 10^8) \qquad (3.15)$$

is fulfilled. The transition between turbulent regime and fast fluidization is observed at the critical velocity U_{se}. At this point the solids begin to be entrained significantly [71].

$$Re_{se} = 1.53 Ar^{0.50} \qquad (2 < Ar < 4 \cdot 10^6) \qquad (3.16)$$

For Geldart D particles, when U_{se} predicted from Equation (3.16) is less than the terminal velocity U_t, U_{se} should be taken as U_t [65].

Grace [68] has suggested a flow regime map in which the different regimes are clearly indicated (Figure 3.8).

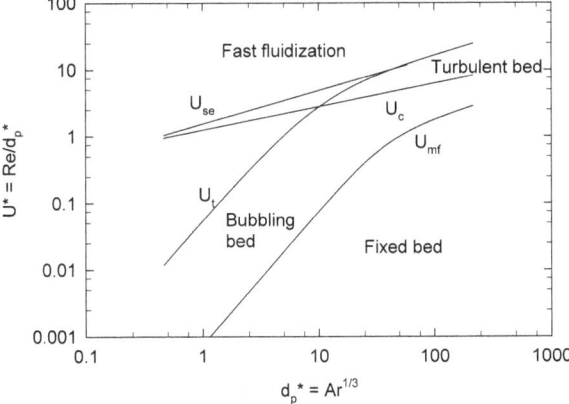

Figure 3.8: Flow regime map suggested by Grace [68]. The x-axis describes the effect of particle characteristics on the fluidization regime whereas the y-axis accounts for gas flow characteristics. The course of U_c and U_{se} are extrapolated beyond their application limits ($Ar < 2$).

3.5. Technical background for the design of chemical looping combustors

For the design of chemical looping combustors, different aspects have to be considered. As already mentioned, fluidized beds offer some optimal features, such as excellent gas-solids contact and the possibility for solids transport, for chemical looping processes. In this study, the design focuses on a dual circulating fluidized bed reactor system. For the operation of a chemical looping reactor system, solids circulation is implicitly necessary at all times. Without solids circulation, no O_2 is transported from the air reactor to the fuel reactor which will consequently lead to the shut down of the process.

The oxygen carrier has a high influence on the design of the reactor system and should therefore be defined prior to the basic and detailed engineering. The oxygen transport capacity R_0 defines the amount of O_2 that can be transported by the carrier and is determined via

$$R_0 = \frac{m_{OC,ox} - m_{OC,red}}{m_{OC,ox}} \quad (3.17)$$

with the masses of the fully oxidized and reduced oxygen carrier $m_{OC,ox}$ and $m_{OC,red}$, respectively. R_0 is highly dependent on the redox system (see Figure 2.2 on page 20) and the amount of inerts in the particle. The higher R_0 the less solids circulation is necessary to run a chemical looping combustor. The actual oxidation state of the particle, henceforth termed solids conversion, can be calculated from the actual mass of the oxygen carrier at a certain position in the reactor system:

$$X_S = \frac{m_{OC} - m_{OC,ox} \cdot (1 - R_0)}{m_{OC,ox} \cdot R_0} \quad (0 \leq X_S \leq 1) \quad (3.18)$$

Theoretically, the oxygen carrier could be fully oxidized in the air reactor and then fully reduced in the fuel reactor, exploiting the maximum oxygen transport capacity of the particle. On the one side, this would minimize the required solids circulation rate in the reactor system. The required solids inventory would be indefinitely high, on the other side. This results from the fact that in this case of an ideally mixed air and fuel reactors, the active sites in each reactor

$$n_{active,AR} = const \cdot (1.0 - X_{AR}) \quad (3.19)$$

$$n_{active,FR} = const \cdot X_{FR} \quad (3.20)$$

approach 0 and therefore there are no active sites for fuel oxidation in the fuel reactor and for O_2 uptake in the air reactor respectively. Further, the required solids inventory will go ad infinitum in this case.

The mean solids conversion in the reactor system $\overline{X}_S$

$$\overline{X}_S = \frac{m_{OC,AR} \cdot X_{AR} + m_{OC,FR} \cdot X_{FR}}{m_{OC,AR} + m_{OC,FR}} \quad (3.21)$$

derives from the distribution of solids between the two reactors and from the kinetic parameters of the oxygen carriers toward O_2 and fuel. When referring to the solids distribution between the two reactors, one could also speak of the solids residence time distribution. These two terms have the same significance which can easily be seen from the following expressions:

$$\tau_{S,AR} = m_{OC,AR}/\dot{m}_{OC} \quad (3.22)$$
$$\tau_{S,FR} = m_{OC,FR}/\dot{m}_{OC} \quad (3.23)$$

and thus

$$\frac{\tau_{S,AR}}{\tau_{S,FR}} = \frac{m_{OC,AR}}{m_{OC,FR}}. \quad (3.24)$$

The solids inventory in each reactor is only dependent on reactor dimensioning and the hydrodynamic conditions in each reactor. The required solids circulation rate for the operation of a chemical looping combustor is calculated from the fuel flow and the oxygen requirement for full oxidation of the fuel:

$$\dot{m}_{OC} = \frac{O_{min} \cdot \dot{m}_{fuel}}{R_0 \cdot \Delta X} \quad (3.25)$$

with

$$\dot{m}_{fuel} = \frac{P_{th}}{LHV} \quad (3.26)$$

and the O_2 demand for full oxidation of the fuel O_{min}. The amount of air that has to be delivered to the air reactor is calculated via

$$\dot{m}_{air} = \lambda \cdot \dot{m}_{fuel} \cdot O_{min} \quad (3.27)$$

with the global air/fuel ratio λ. From Equations (3.19), (3.20), (3.21) and (3.25) one can derive a relation between the active sites in the reactor system and the amount of circulating solids (with some simplifications [5] and constants c_1 and c_2):

$$n_{active} = c_1 \cdot \left(m_{OC,AR} \cdot (1 - \overline{X}_S - \frac{c_2}{\dot{m}_{OC}}) + m_{OC,FR} \cdot (\overline{X}_S - \frac{c_2}{\dot{m}_{OC}}) \right) \quad (3.28)$$

Optimization of this relation toward $\dot{m}_{OC}$ results in the expression

$$\frac{m_{OC,AR}}{\dot{m}_{OC}^2} + \frac{m_{OC,FR}}{\dot{m}_{OC}^2} = 0 \quad (3.29)$$

which has its optimum at infinitely high solids circulation rate and $\triangle X_S = 0$, respectively. In practice, the difference in solids conversion between air and fuel reactors is in the range of 10 % which corresponds to $\triangle X = 0.10$. The mass of oxygen carrier in both reactors is given by the pressure drops in the reactors, i.e.

$$m_{AR} = \frac{\triangle p_{AR} \cdot A_{AR}}{g} \quad (3.30)$$

and

$$m_{FR} = \frac{\triangle p_{FR} \cdot A_{FR}}{g}. \quad (3.31)$$

3.6. Design of a 120 kW CLC pilot rig

In 2006, a chemical looping combustion pilot rig has been designed at Vienna University of Technology and has been erected in 2007. Figure 3.9 shows the basic arrangement of the CLC reactor system with all major auxiliary units. Air (optionally preheated) is used as fluidization agent for the air reactor and is introduced via nozzles at two different heights. The air reactor exhaust gas is cooled down to 300−400 °C and a small fraction of the gas is directed to gas analyzers (CO, CO_2 and O_2 content analyzers). Gaseous fuel is used as fluidization agent for the fuel reactor but is introduced only at one level (lowest point). The fuel reactor exhaust gas is also cooled down to 300 − 400 °C and partially directed to gas analyzers (CO, CO_2, O_2, H_2, CH_4, N_2, Ar and higher hydrocarbons content analyzers).

After cooling of the air reactor and fuel reactor exhaust gases, both streams are mixed and directed to a post combustion chamber. This way, all combustibles that are still present in the gas stream are oxidized. This is specially important in case of CLR operation when the fuel reactor offgas consists mainly of H_2 and CO. The post combustor

---

[5] The effect of solids circulation rate on the reactivity of the particles, as described in section 5.4 and by Garcia-Labiano et al. [47], is neglected. Considering this aspect, the relation would look somewhat different but has the same trend.

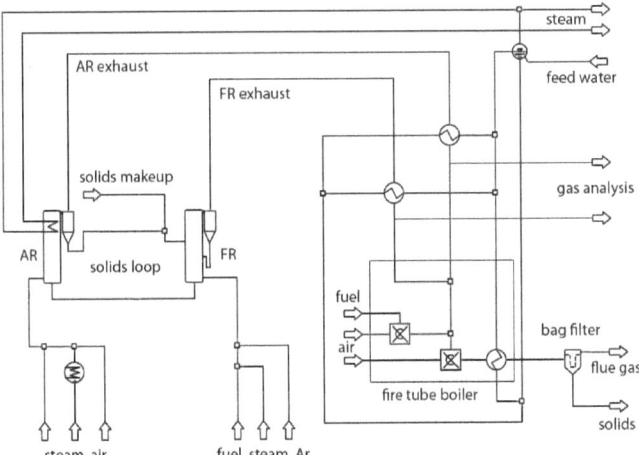

Figure 3.9: Arrangement of the CLC reactor system and major auxiliary units. This process flow diagram results from process modeling. For simplicity reasons, the steam for loop seal fluidization and argon for pressure tab flushing are added to air and fuel.

is designed as fire tube boiler which cools the offgas to a temperature of approx. 200 °C. A bag filter just prior to the stack avoids the emission of solids to the environment.

3.6.1 Reactor system

The unit applies the DCFB reactor concept with a fast fluidized air reactor and a turbulent fuel reactor. The boarder limits between the different flow patterns in gas-solids fluidized beds are described by Equations (3.15) and (3.16). Since the pilot rig is also designed for chemical looping reforming, some compromises in the design have to be made.

The CLC pilot rig is primarily designed for a Ni-based oxygen carrier with $NiAl_2O_4$ support. The main characteristics of this oxygen carrier are shown in Table 3.1. All loop seals, however, are designed to support very high solids circulation rates to allow the operation with oxygen carriers with much lower oxygen transport capacity. The standard operating cases for chemical looping combustion and chemical looping reforming, respectively, are shown in Table 3.2.

The superficial velocity in the air reactor is chosen well above U_{se} to assure proper

parameter	unit	value
active metal		Ni/NiO
support material		$NiAl_2O_4$
active Ni content	%	40
oxygen transport capacity	kg/kg	0.084
mean particle diameter	mm	0.120
apparent density ($X_S = 1.0$)	kg/m^3	3200
sphericity		0.99

Table 3.1: Oxygen carrier for the CLC pilot rig

solids circulation in all operating cases. This is especially important for CLR operation where the superficial velocity in the air reactor is rather low ($\lambda \ll 1.0$). In some operating cases, the air reactor fluidization drops into the turbulent regime for this reason. The nominal fuel power for CLR operation is increased to 200 kW which also increases the amount of air in the air reactor (and thus superficial velocity).

Unfortunately, the reactor height is limited by the surrounding laboratory to values far below industrial standard. The air reactor height is limited with 4.1 m whereas the fuel reactor height is limited with 3.0 m. With this height, the gas residence time in the fuel reactor is very short when operated in the turbulent regime. Therefore, the superficial velocity in the fuel reactor is set to a value slightly below turbulent regime but well above the terminal velocity U_t. Figure 3.10 shows a regime map in which different operating cases of the reactor system are sketched.

		CLC operation		CLR operation	
parameter	unit	AR	FR	AR	FR
inlet gas flow	m$_N$/h	138.0	12.0	75.7	20.0
outlet gas flow	m$_N$/h	113.9	35.9	59.8	59.9
temperature	K	1213	1123	1223	1123
fluid		N_2,O_2	H_2O,CO_2	N_2^*	H_2,H_2O,CO,CO_2
Archimedes number		7.55	9.13	8.46	5.47
superficial velocity	m/s	7.32	2.08	4.91	3.45
U/U_{mf}		1280	315	756	401
U/U_t		15.5	3.8	9.2	4.8
fuel power (natural gas)	kW	120		200	
LHV of fuel	MJ/kg	48.8		48.8	
air/fuel ratio		1.2		0.5	

*The O_2 should be converted in the lower section of the air reactor.

Table 3.2: Design parameters of the chemical looping pilot rig for CLC and CLR operation

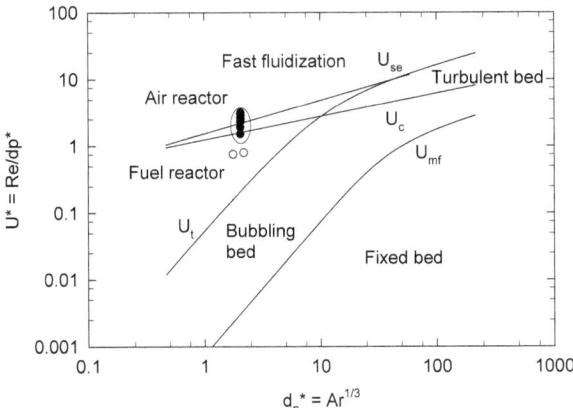

Figure 3.10: Operating conditions in air and fuel reactors shown in a flow regime map. The air reactor is operated in fast fluidization regime but drops into the turbulent regime in some CLR operating cases. The fuel reactor is operated below the turbulent regime to assure a minimum gas residence time in the reactor.

For effective control of the solids circulation rate, the possibility of air staging is included in the design concept. The shift of a fraction of the total air to a higher level results in a reduction of the global solids circulation rate. This has already been demonstrated in the cold flow model (see Figures 3.4 and 3.5(a)). The cyclone separators for particle separation from the air reactor and fuel reactor offgases are designed according to Hugi [72]. A summary of the major reactor dimensions is shown in Table 3.3.

In total, three loop seals are implemented in the reactor system. These are

- the upper loop seal in the connection between air reactor downcomer and fuel reactor,
- the lower loop seal in the bottom part of the reactor system which connects air and fuel reactors and
- the internal loop seal for the internal circulation in the fuel reactor.

The upper and lower loop seal have identical dimensions (see Table 3.3). The internal loop seal is somewhat smaller since the internal solids circulation rate is expected to be much smaller than the global solids circulation rate. All loop seals are fluidized with steam and the ratio of U/U_{mf} is kept as low as possible to ensure proper gas sealing.

parameter	unit	AR	FR
reactor diameter	m	0.150	0.1593
reactor height	m	4.1	3.0
height of prim. air/fuel inlet	m	0.025	0.060
height of sec. air/fuel inlet	m	1.325	—
cyclone separator diameter	m	0.310	0.260
upper loop seal cross section	m^2	9.9E-3	
lower loop seal cross section	m^2	9.9E-3	
internal loop seal cross section	m^2	6.4E-3	

Table 3.3: Dimensions of major reactor components

3.6.2 Reactor cooling system

The cooling duty of a CLC reactor system to maintain stable operation temperature is shown in Figure 2.6(a) on page 25. At typical global air/fuel ratios of combustors ($\lambda = 1.0 - 1.3$), a great share of the total heat has to be extracted directly from the reactor system. In principle, there are three approaches for this cooling:

1. The bed material can directly be cooled in a fluidized bed heat exchanger. This method is applied in most modern CFB boilers.

2. Direct cooling of the reactor walls allows heat extraction from air and/or fuel reactor.

3. The global air/fuel ratio can be increased to a value that no additional heat extraction is necessary. With this approach, however, global solids circulation rate, air/fuel ratio and reactor temperature cannot be controlled independently from each other.

In the present design, the second approach has been applied. The air reactor is equipped with three different cooling jackets which can be streamed with different fluids. The uppermost cooling jacket is cooled with air which is heated up to 500 °C (depending on operating conditions). A quench cooler cools the hot air down to approx. 200 °C. The air is then mixed with the steam in the steam drum and optionally directed to the lowest cooling jacket for further reactor cooling or directly to the stack. The amount of air for the uppermost cooling jacket and the amount of the steam/air mixture to the lowest cooling jacket (and thus cooling duty) can be controlled with valves. The third cooling jacket is designed as evaporator and can optionally be connected or disconnected on the water side prior to but not during the experiments (thermal shock of the air reactor wall).

All other heat exchangers in the pilot rig are connected to a steam drum and operated with boiling water at atmospheric pressure. The air reactor and fuel reactor exhaust gas coolers are designed as tube in tube heat exchangers. The post combustor is designed as a two pass fire tube boiler. Cooling water circulation is accomplished with natural circulation. The produced steam from all heat exchangers is collected in the steam drum, used for cooling of the air reactor and then directed to the stack. For simplicity reasons, a condenser of the steam is not included in the pilot rig arrangement.

3.6.3 Auxiliary units

Different auxiliary units are necessary to operate the pilot rig. Gas compressors for air and natural gas are necessary to overcome the pressure drops in the system. H_2, CO and C_3H_8 are considered as alternative fuels and are provided from pressurized gas cylinders. Heat exchangers for air reactor and fuel reactor offgases cool the hot gases to an intermediate state at which samples for gas analysis can be taken. A post combustor with integrated gas cooling ensures complete oxidation of all combustibles and cools the offgas to approx. 200 °C. The subsequent bag filter ensures maximum particle separation from the gas stream before the flue gas is directed to the stack. A steam drum is used for cooling water distribution between the different heat exchangers.

A programmable logic controller (PLC) in combination with a LabView application is used for process control and data logging of all relevant data. For each stable operation point, mass and energy balances are performed with the IPSEPro simulation tool. Owing to the high number of different measurements (flows, temperatures, pressures, gas compositions, ...), the set of equations is overdetermined. This is used to minimize the uncertainties of all measured values. More details on this evaluation are presented by Bolhar-Nordenkampf et al. [73].

4

OPERATING RESULTS OF THE CLC PILOT RIG

4.1. Hydrodynamic operation of the DCFB reactor system

This section summarizes the hydrodynamic operating performance of the 120 kW dual circulating fluidized bed pilot rig for chemical looping combustion in hot operation. For this purpose, air and fuel reactor pressure profiles at different operating conditions are discussed. Further, the active solids inventory in the reactor system and the impact of air staging on the global solids circulation rate and the hydrodynamic profile of the fuel reactor are determined and presented.

4.1.1 Pressure profiles of air and fuel reactors

In total, the CLC pilot rig is equipped with approx. 30 pressure tubs. These are distributed all over the pilot rig, including reactor system, heat exchangers and other auxiliary units. 18 pressure measurements are positioned in the DCFB reactor system. These allow the determination of the pressure profiles such as shown in Figure 4.1.

The lower loop seal (LLS) is the lowest point in the reactor system and thus is exposed to the highest pressure. The connections between LLS and air reactor and fuel reactor respectively, are fluidized with steam as well and operate in the bubbling regime. Since the void fraction is rather low at this point, a steep pressure decrease with increasing height is observed. The nearly linear pressure profile in the air reactor indicates an almost constant void fraction along the reactor height. Only in the bottom part of the air reactor, the void fraction is slightly decreased (denser bottom bed). The fuel reactor, on the other hand, is characterized by a much denser bottom bed. Most of the total pressure drop in the fuel reactor occurs between the first two measuring points. This indicates a dense bottom bed with much fewer particles in the higher sections of the reactor and much lower solids circulation compared with the air reactor.

The solids entrainment of a reactor can be evaluated by determining the pressure loss in the respective loop seal. The two-phase flow (gas-solids) in the loop seals generates a pressure loss due to friction. This means, the higher the pressure drop in the loop seal, the higher is the solids circulation rate in the respective reactor (assuming equal

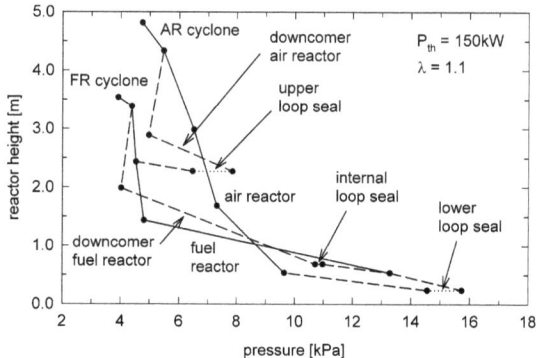

Figure 4.1: Pressure profile of the DCFB reactor system at 150 kW fuel power

fluidization and loop seal dimensions). From this point of view, the internal solids circulation rate is much lower than the global solids circulation rate (see Figure 4.1). A pressure drop vs. mass flow diagram for the upper and lower loop seal is shown in Figure 4.2. For both loop seals a linear trend is observed. In general, the upper loop seal (ULS) experiences a higher pressure difference at equal solids mass flow. This is probably caused by the slight differences in fluidization. [1]

Air staging is used to control the global solids circulation rate in the reactor system. This measure will mainly change the hydrodynamic behavior of the air reactor but might also have some influence on the fuel reactor. To investigate this influence, pressure profiles are determined at three different air staging positions. The results are plotted in Figure 4.3. Significant changes in the bottom bed of the air reactor are observed. At lower primary fluidization, the pressure difference in the bottom bed is increased which indicates a denser and/or higher bottom bed. Further, air staging results in an increased air reactor solids inventory. The pressure profile in the upper region is almost identical in all cases. The influence of air staging on the fuel reactor is very low. From the obtained results, air staging does not show any potential problems for the operation of the CLC pilot rig.

The pilot rig is designed for a wide range of different fuel powers. Especially in the case of chemical looping reforming, an increased fuel power is necessary to assure sufficient

---

[1] i.e. different amount of steam and back pressure.

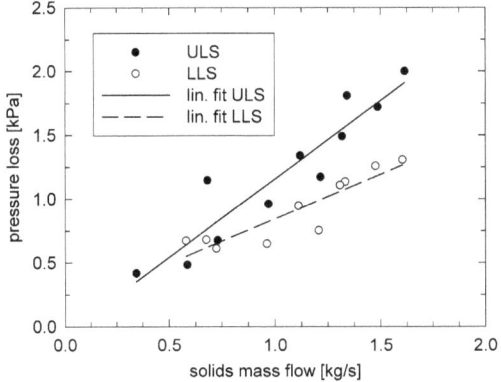

Figure 4.2: Pressure loss in the upper and lower loop seals as a function of solids mass flow. Both loop seals have identical dimensions and very similar fluidization. The back pressure in the ULS is approx. 10 kPa lower than in the LLS (depending on operating conditions).

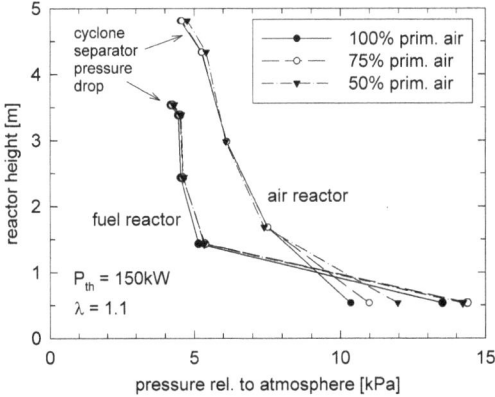

Figure 4.3: Effect of air staging on the pressure profile in air and fuel reactors

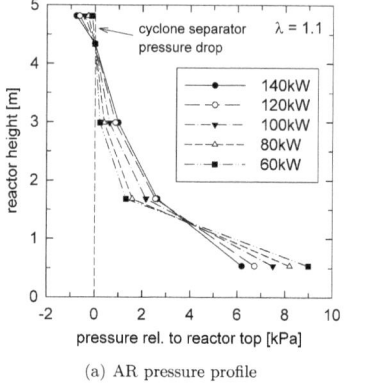
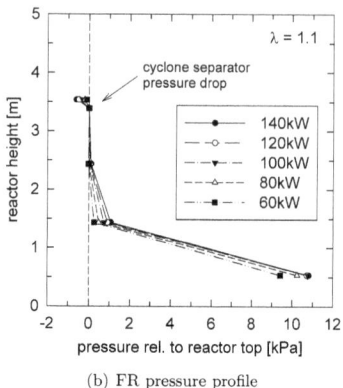

(a) AR pressure profile (b) FR pressure profile

Figure 4.4: Effect of fuel power variation on the pressure profiles in air and fuel reactors

solids circulation.[2] Therefore, the influence of the fuel power on the hydrodynamics of the reactor system is investigated and shown in Figure 4.4. Since the fluidization of the air and fuel reactors is accomplished with the introduced air and fuel respectively and no gas recirculation is considered, this plot directly shows the impact of a change in the fluidization numbers in both reactors. In the air reactor (Figure 4.4(a)) a denser bed (higher dense bed height and/or lower void fraction) in the bottom region is observed when the fluidization is decreased. This is mainly caused by the lower solids circulation rate at lower fuel power. At high air reactor fluidization, the solids are rapidly transported back to the fuel reactor which decreases the solids residence time in the air reactor. The fuel power change also shows some impact on the fuel reactor pressure profile (Figure 4.4(b)). At decreased fuel power, and thus fluidization, the amount of solids in the upper regions of the reactor is decreased.

When increasing the fuel power, some of the air reactor solids inventory is shifted to the fuel reactor. In the range of 60 – 100 kW the air reactor loss and fuel reactor gain are almost equal. At higher fuel power, however, the fuel reactor solids inventory remains constant whereas the air reactor inventory is reduced further. It is assumed that, owing to the increased solids circulation, most of this missing material is present in the air reactor cyclone separator and downcomer. Some of this material is certainly also present in the internal loop. Compared with the global solids loop, however, this

---

[2]This measure also increases the air flow in the air reactor which enhances solids circulation.

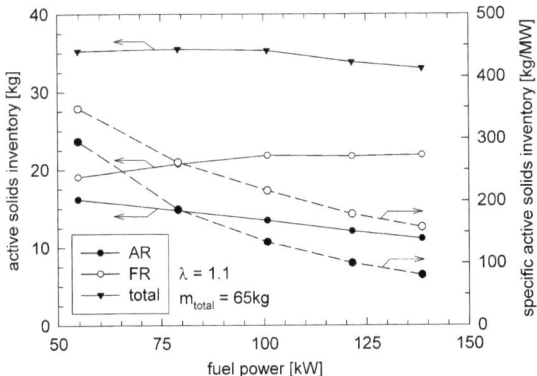

Figure 4.5: Effect of power variation on the active solids in air and fuel reactors

fraction should be rather low.

The total solids inventory of the CLC pilot rig to operate with reasonable pressure drops in air and fuel reactors is in the range of $55 - 80\,\text{kg}$.[3] Only a fraction of this amount, however, is present in both reactors where the chemical reactions proceed. The rest can be found in the loop seals, downcomers and cyclone separators.[4] The active amount of material in each reactor, henceforth termed active solids inventory, has been defined in Equations (3.30) and (3.31) and is derived from the pressure drops in air and fuel reactors, respectively.

The active solids inventory in both reactors is also influenced by the fluidization of the reactors. This is shown in Figure 4.5 using the example of a fuel power variation. At higher fuel power, a slight reduction of the total active solids inventory is observed. Since the loop seal inventory should be nearly constant for different fuel powers, the missing solids must be present in the downcomers and cyclone separators. This assumption is supported by the observed increased solids circulation rate at higher air reactor fluidization (see Figure 4.6). Further, a shift of active solids inventory from the air reactor to the fuel reactor is observed. This is probably caused by the increased back pressure in the air reactor compared with the fuel reactor.

Beside the active solids inventory, the specific active solids inventory in kg/MW is

---

[3] This value corresponds to fully oxidized particles.
[4] Some gas conversion will certainly also take place in the cyclone separators. Owing to the lack of a possibility to measure this quantity, it is neglected for the calculation of the active solids inventory.

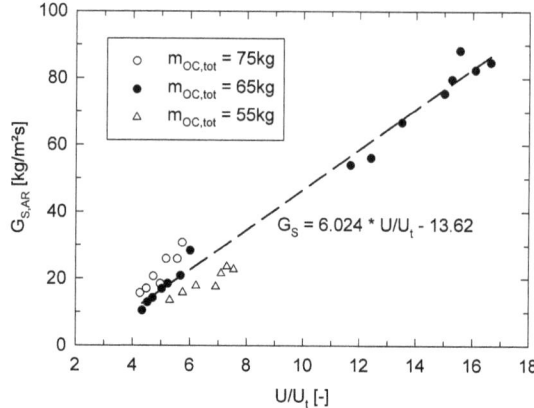

Figure 4.6: Global solids circulation rate $G_{S,AR}$ as a function of total solids inventory and fluidization number U/U_t

plotted in Figure 4.5. Naturally, the specific active solids inventory decreases with fuel power. Since this value also affects the gas conversion, it will have to be optimized in an industrial application.

The global solids circulation rate can be expressed in terms of the air reactor net solids flux $G_{S,AR}$, defined as

$$G_{S,AR} = \frac{\dot{m}_{OC}}{A_{AR}} \quad (4.1)$$

and can be determined by measuring the solids conversion of the particles at the fuel reactor inlet and outlet (Equation (3.25)). As shown in Figure 4.6, $G_{S,AR}$ is mainly influenced by the degree of fluidization in the air reactor. A linear trend of $G_{S,AR}$ with U/U_t is observed in the investigated operation range. The total solids inventory has a minor impact on the results. This plot shows one of the problems when operating the pilot rig at very low global air/fuel ratios (CLR). Owing to the low fluidization number (in this case U/U_t), the solids circulation rate is significantly reduced which results in a high difference in solids conversion of the particle and a high temperature difference between air and fuel reactors.[5]

---

[5] To transfer the necessary heat for the reforming reactions in the fuel reactor, a temperature difference between air and fuel reactors will emerge to fulfill the energy balance.

4.2. CLC performance with different oxygen carriers

The 120 kW pilot rig for CLC at Vienna University of Technology is designed for a Ni-based oxygen carrier. For commissioning of the pilot rig, ilmenite, a natural mineral with the notation $FeTiO_3$, is applied for safety reasons. During this test campaign, different results are obtained. After commissioning and optimization of the pilot rig, most experiments are performed with the two Ni-based particles (OC-A and OC-B) specified in Table 4.1. This specification is only slightly different from the specification the pilot rig was originally designed for (see Table 3.1 on page 46).

4.2.1 Experimental procedure and evaluation of results

For the evaluation of the results, it is advantageous to introduce different parameters. In the case of CH_4 combustion, the CH_4 conversion X_{CH_4} is calculated from

$$X_{CH_4} = 1 - \frac{x_{CH_4}}{x_{CH_4} + x_{CO_2} + x_{CO}} \qquad (4.2)$$

where x_i is the volume fraction of the species i in the fuel reactor exhaust gas. In the same way, the CO conversion is calculated for the case of syngas combustion (CO and H_2 as fuel):

$$X_{CO} = 1 - \frac{x_{CO}}{x_{CO_2} + x_{CO}} \qquad (4.3)$$

In the case of H_2 as fuel, the situation is somewhat more complicated. Since the gas analysis can analyze dry gases only, an inert gas has to be added to the fuel to determine X_{H_2}.[6] For this purpose, N_2 is added to the fuel and X_{H_2} reads as

$$X_{H_2} = \frac{\dot{n}_{H_2,in} - \dot{n}_{H_2,out}}{\dot{n}_{H_2,in}} = 1 - \frac{\dot{n}_{N_2,in} \cdot x_{H_2}}{\dot{n}_{H_2,in} \cdot x_{N_2}}. \qquad (4.4)$$

parameter	unit	OC-A	OC-B
active metal		Ni/NiO	Ni/NiO
support material		$NiAl_2O_4$	$NiAl_2O_4/MgAl_2O_4$
NiO content	%	≈ 40	≈ 40
oxygen transport capacity	kg/kg	0.08568	0.08844
d_p	mm	0.120	0.120

Table 4.1: Ni-based oxygen carriers applied in the CLC pilot rig

---

[6] When operating with pure H_2 as fuel, the dry exhaust gas will consist of H_2 only, independent of fuel conversion. By adding an inert gas, an element balance of this substance can be applied to calculate the H_2 conversion.

In the case of syngas combustion (mixtures of CO and H_2), N_2 does not have to be added to the fuel and the H_2 conversion can be calculated with the aid of a C-balance. In this case, X_{H_2} reads as

$$X_{H_2} = 1 - \frac{(\dot{n}_{CO,in} + \dot{n}_{CO_2,in}) \cdot x_{H_2}}{\dot{n}_{H_2,in} \cdot (x_{CO} + x_{CO_2})}. \tag{4.5}$$

The CO_2 yield describes the selectivity of the reactions toward CO_2

$$\gamma_{CO_2} = \frac{x_{CO_2}}{x_{CH_4} + x_{CO_2} + x_{CO}} \tag{4.6}$$

and is thermodynamically limited to values slightly below 1.0 when Ni-based OCs are used [73]. In Equations (4.2), (4.3), (4.5) and (4.6) it is assumed that no carbon is formed in the fuel reactor which will be confirmed in section 4.2.4.

The CLC pilot rig offers the possibility of solids sampling during hot operation. Therefore, the solids conversion can be determined. Using this value, the solids circulation rate can be calculated via Equation (3.25). The set-up for solids sampling is shown in Figure 4.7. A heat-resistant pipe is introduced directly into the fluidized bed in the downward direction (A). In this way, solids are prevented from advancing toward the instrumentation during times when no solids are extracted. The pipe is then bent into a vertical direction. A ball valve (B) which does not distract the solids path in the fully

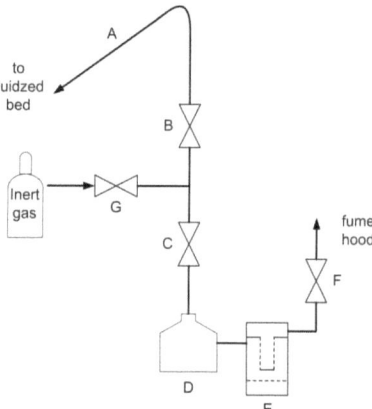

Figure 4.7: Experimental set-up for solids sampling from fluidized beds

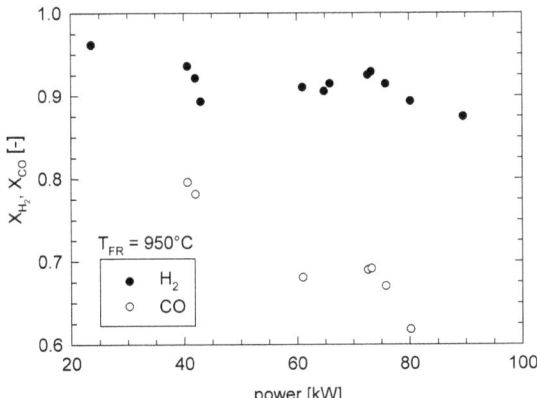

Figure 4.8: H_2 and CO conversion using ilmenite as oxygen carrier

open position seals the connection to the fluidized bed at this point. From there, the solids pipe is connected to the solids deposit (D) with another ball valve (C) in between. The solids deposit outlet is equipped with a filtering device (E) and another valve (F). Between the two solids valves (B) and (C) a connection for inert gases with another valve (G) is attached. In the case of hot sampling, the solids container (D) is cooled with cold water to avoid overheating of the instrumentation. More details on the solids sampling procedure and evaluation of the samples are presented in Paper IV.

4.2.2 Performance of ilmenite for chemical looping combustion

Prior to the operation with the designed Ni-based oxygen carrier, a series of experiments with ilmenite ($FeTiO_3$) as oxygen carrier is performed. For this purpose the CLC pilot rig is fueled with H_2 and CO. Unfortunately, the used particles have a relatively large mean diameter ($200 - 300\,\mu m$).

Figure 4.8 shows the results of this first investigation. The H_2 conversion is in the range of $0.90 - 0.95$ and decreases slightly with fuel power. This decrease is much more pronounced in the case of CO conversion. X_{CO} reaches values up to 0.80 at $40\,kW$ fuel power but is reduced to $0.60 - 0.65$ at $80\,kW$. These results, however, are to be regarded as first results with ilmenite and have a high potential for optimization.

Beside an increase of the reactor height[7] and better suited particle size distribution, the optimization of different operating parameters of the pilot rig should increase the gas conversion. Nevertheless, these results are very promising and reveal ilmenite as a potential oxygen carrier for H_2 and CO rich gases (e.g. from coal gasification with steam). More results obtained with ilmenite and other natural minerals as oxygen carrier are presented by Pröll et al. [74].

4.2.3 Performance of Ni-based particles

After this first series of experiments with ilmenite, two different Ni-based particles (see Table 4.1) are tested. The first oxygen carrier (OC-A) is used for H_2, CO and CH_4 combustion. Unfortunately, only a limited amount of the second oxygen carrier (OC-B) is available for testing. Therefore, OC-B is mixed with OC-A in a ratio of 1 : 1. This mixture, henceforth termed OC-AB, is used for CH_4 combustion.

Combustion of H_2 and CO

Compared to ilmenite, the tested Ni-based oxygen carriers have a much higher reactivity. Figure 4.9 clearly shows that the H_2 conversion is very close to the thermodynamic maximum. Only in case of very low total solids inventory (55 kg), distinctive differences from maximum conversion are observed. CO is also converted to a high degree but the conversion is somewhat lower compared with H_2. In Figure 4.9 the CO conversion at 65 kg total solids inventory is higher compared with 75 kg. This is mostly owing to the fact that the latter experiments were performed at an average fuel reactor temperature of approx. 30 K below the former ones. This indicates that, in this case, the operating temperature has a much higher impact on the fuel conversion than the total solids inventory.

Dueso et al. [75] have reported that the water gas shift reaction (WGS)

$$CO + H_2O \rightleftharpoons CO_2 + H_2 \qquad (4.7)$$

plays an important role in the conversion of CO when Ni-based oxygen carriers are used. The WGS reaction is supposed to proceed much faster than the combustion of CO with NiO. Therefore, much of the present CO is converted to CO_2 via the WGS reaction with subsequent oxidation of the produced H_2 at the oxygen carrier.

---
[7]The air and fuel reactor height is limited to 4.1 m and 3.0 m respectively owing to the surrounding laboratory height. The height of a CFB, however, is not scalable and thus should be much larger in the CLC pilot rig. This would increase the gas residence time in air and fuel reactors and thus gas conversion.

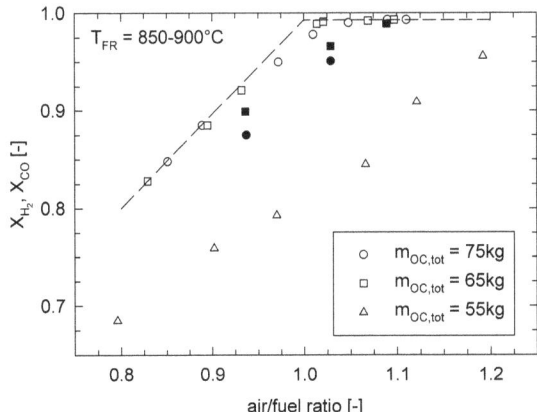

Figure 4.9: H_2 (o) and CO (•) conversion using a Ni-based oxygen carrier (OC-A) at a fuel power of 60 – 80 kW

The evaluation of solids samples from the experiments reported in Figure 4.9 is shown in Figure 4.10. The solids conversion in the fuel reactor is very low in most conditions. Only in the case $\lambda > 1.05$ and fairly high total solids inventory (65 – 75 kg), values above 0.2 are observed. The solids conversion at 65 kg total solids inventory is clearly higher compared with 75 kg at $\lambda > 1.0$. This result is rather surprising but, as already mentioned, the experiments at 65 kg total solids inventory are performed at approx. 30 K increased temperature. The temperature dependance of the reactions which has also been observed in CO conversion (see Figure 4.9), might be one reason for the observed difference in solids conversion.

Combustion of CH_4

The H_2 fueled experiments have clearly shown the very high reactivity of OC-A. In a new series of experiments, the CLC pilot rig is fueled with CH_4 at a fuel power of approx. 140 kW. Both, OC-A and OC-AB are used.

A temperature variation for CH_4 combustion in the CLC pilot rig is shown in Figure 4.11. OC-A shows high CH_4 conversion (up to 0.95) which is nearly independent of the fuel reactor temperature. The CO_2 yield, on the other hand, is highly dependent on the reactor temperature. A maximum value of 0.89 is observed. OC-AB shows an

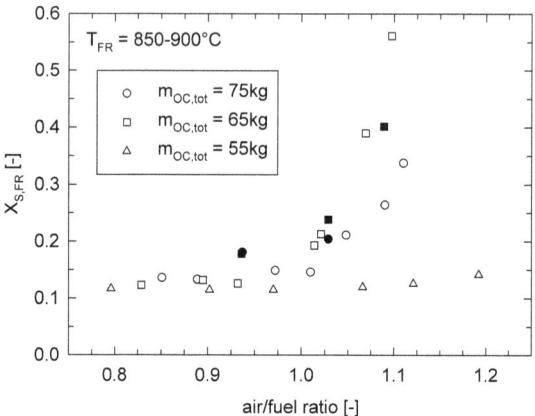

Figure 4.10: Solids conversion in the fuel reactor for H_2 (○) and CO (●) combustion using a Ni-based oxygen carrier (OC-A) at a fuel power of 60 – 80 kW

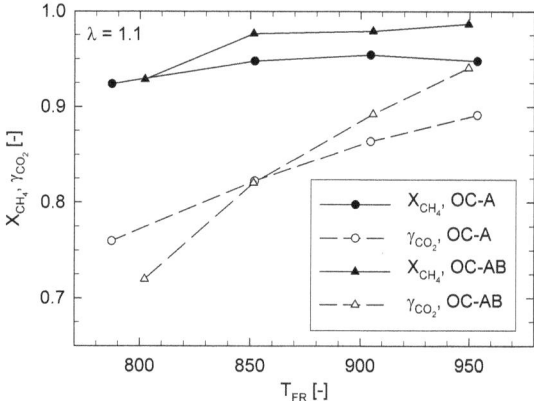

Figure 4.11: Effect of temperature on the CH_4 conversion and CO_2 yield for OC-A and OC-AB

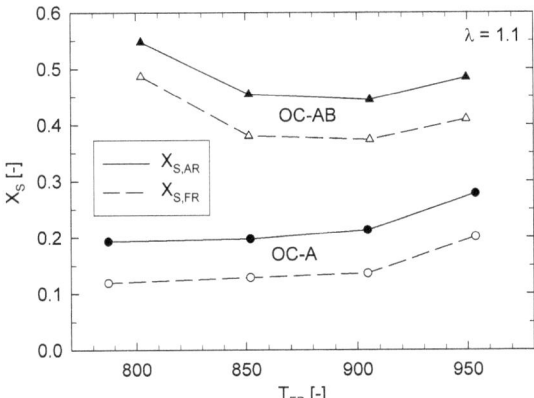

Figure 4.12: Effect of temperature on the solids conversion in the air and fuel reactors for OC-A and OC-AB

even better performance than OC-A. Both, CH_4 conversion and CO_2 yield, reach higher values than OC-A (up to 0.99 and 0.94 respectively). The CH_4 conversion is nearly constant at high fuel reactor temperature but below 850 °C a decrease is observed. The temperature dependance of the CO_2 yield is even more distinctive than for OC-A.

The evaluation of the sampled solids during operation of the just discussed experiments is shown in Figure 4.12. OC-A is almost fully reduced in all cases.[8] It could be concluded that the increased solids conversion of OC-AB results in an increased CO_2 yield. This effect, however, has to be investigated more in detail in the future. In the case of identical experimental set-up, the solids oxidation of a particle can be increased by two means:

1. Decreased speed of reaction in the fuel reactor and
2. increased speed of reaction in the air reactor.

Since higher fuel conversion is observed for OC-AB, it has to be concluded that the MgO content somehow enhances the reactivity of the particles toward O_2.

The influence of the global air/fuel ratio on the performance of CH_4 combustion with OC-A and OC-AB is investigated in a next series of experiments. In Figure 4.13, OC-A

---

[8]The solids conversion considers the conversion of all of the active metal. Some of the active metal, however, has to be considered as inactive owing to the possibility of being trapped somewhere in the inner structure of the particle. Therefore, $X_S = 0.1$ might already be the minimum solids conversion of the particle.

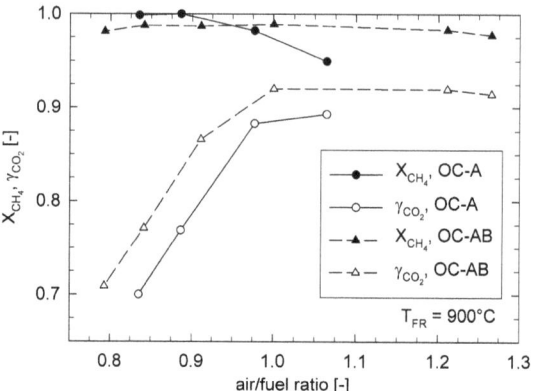

Figure 4.13: Effect of air/fuel ratio variation on the CH_4 conversion and CO_2 yield for OC-A and OC-AB

shows complete CH_4 conversion for $\lambda < 0.9$ but a significant CH_4 slip at higher air/fuel ratios. OC-AB, on the other hand, does not show this dependance at all; the CH_4 conversion is nearly constant at 0.985 in all cases. OC-AB also shows an increased CO_2 yield with a maximum value of 0.92. Naturally, the CO_2 yield decreases strongly when the air/fuel ratio is decreased below 1.0.

The influence of the air/fuel ratio on the solids conversion is shown in Figure 4.14. Just as in the case of the temperature variation, the solids conversion of OC-AB is much higher than for OC-A which, again, might be the reason for the increased CO_2 yield. Both oxygen carriers indicate an increase of solids conversion with the global air/fuel ratio but only for OC-AB a steep decrease in solids conversion is observed for $\lambda < 1.0$.

4.2.4 Carbon formation in the fuel reactor

Since carbon formation in the fuel reactor will lead to CO_2 emissions from the air reactor and thus results in a decreased CO_2 capture efficiency, it has been discussed as a problem in CLC. The amount of carbon produced in the fuel reactor can be derived from CO_2 measurements in the air reactor exhaust gas.

In all experiments performed in the CLC pilot rig, no additional steam is added to the fuel. Some steam, however, is available from the fluidization of the loop seals. In the highest case, this amount would correspond to a H_2O/CH_4 molar ratio of approx.

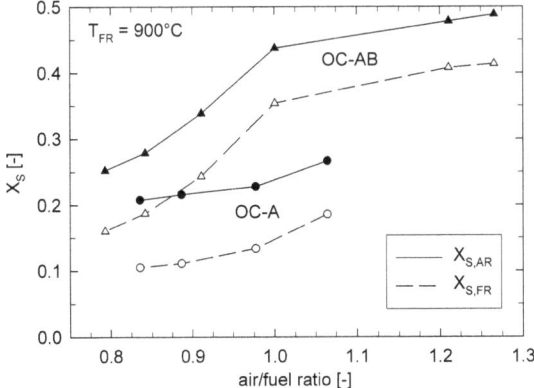

Figure 4.14: Effect of global air/fuel ratio variation on the solids conversion in air and fuel reactors for OC-A and OC-AB

0.4 : 1. Highest case means that most of the steam from the lower loop seal is directed to the fuel reactor. This is, however, very unlikely since the flow of solids in the LLS is directed to the air reactor. Most of the steam from the ULS is directed to the fuel reactor but this is far above the fuel inlet in the reactor where the reforming and combustion take place. Despite this very low H_2O/CH_4 ratio and the low solids conversion of the particles, no carbon formation has been measured in any of the reported operating cases. In experiments not shown here, some carbon formation has been observed at air/fuel ratios below 0.4.

4.3. Summary and outlook

The dual circulating fluidized bed pilot rig has proven very good operating conditions for CLC. The solids circulation rate is sufficiently high and can be controlled very effectively with air staging. In the current design, the solids entrainment and bed expansion of the fuel reactor fluidized bed are quite low and should be increased in the next design. Also, the reactor height is very low and should be increased to at least 12 − 20 m. This would certainly improve the gas conversion in the air and fuel reactors further.

Ilmenite shows adequate H_2 and CO conversion and has a high potential for syngas combustion and CLC of solid fuels. The obtained results, however, have a high potential for optimization and even better results are expected in the future.

Ni-based oxygen carriers show very high reactivity and close to maximum conversion of H_2 and CO. For CH_4 combustion, two different oxygen carriers have been used. With OC-AB, which is a mixture of two different Ni-based particles, nearly full CH_4 conversion and a CO_2 yield of up to 0.95 is observed.

As already mentioned, it would be very interesting to perform further experiments with ilmenite as oxygen carrier. The operation of the CLC pilot rig has been optimized in the recent past and thus even better fuel conversion is expected now. Further, long-time continuous looping testing with OC-A and OC-AB could reveal possible changes in the oxygen carrier and their effect on the performance.

5

MODELING OF THE CLC PILOT RIG

5.1. Introduction Modeling is a widely used tool to qualitatively and quantitatively evaluate the influence of different operating parameters on a process. In technology development, it is a very important tool for better understanding and interpretation of laboratory results and subsequent scale-up. A validation of the used model allows predictions on the performance of similar and scaled-up plants. It has to be kept in mind, however, that modeling always means a compromise between detailed description and a reasonable number of model parameters that can be validated by experimental data.

In CFB technology, much modeling has been performed on solid fuel combustion (e.g. Haider and Linzer [76] for coal and Adanez et al. [77] for biomass combustion; an overview is presented by Basu [78]). These simulations usually focus on the conversion of fuel particles (devolatilization and char combustion), SO_2 and NO_x production as well as heat transfer [78]. In these simulations, the solids inventory acts as heat transfer agent but it is not involved in chemical reactions. Modeling of catalytic gas-solid fluidized bed reactors has been performed by Marmo et al. [79].

Different models have been presented in literature for the description of circulating fluidized bed reactors (see Grace and Lim [80] for an overview). In most cases, like circulating fluidized bed combustion (CFBC), the bed material does not take part in the reaction acting as a reactant. Only the fuel particles react with the gas phase while the bulk sand defines the fluid dynamic regime and enhances heat exchange. In the case of chemical looping, where the bed material undergoes repeated cycles of oxidation and reduction, the bulk bed material is crucially involved in the fuel conversion reactions. As there is no oxidation possible without the presence of bed material, the condition of the bed material (i.e. solids conversion) has to be considered in the model.

5.2. Model development

The applied model consists of a number of different sub-models. These include a reaction model (reaction rates of gases and solids), a fluid dynamic model (interactions of gases and solids) and the set-up of the reactor system which includes the implementation of mass and energy balances. The required thermodynamic properties of gases and solids are calculated according to Burcat and Ruscic [81]. All ideal gas data used in this work are initially based on the JANAF-tables [82].

5.2.1 Model structure

The main focus of the applied model is on the chemical reactions (homogeneous and heterogeneous) taking place in the reactor. Each reactor is divided into a finite number of cells along the height axis (one dimensional model). In each cell, a sub-model performing the chemical reactions dependent on the local fluid dynamics (solids hold-up, solids distribution, etc.) is implemented. The mass balances of all elements except those potentially present as solids (i.e. metals, oxygen) are calculated for each cell. Energy and solids are balanced globally across the whole reactor. This implies that each reactor is modeled as being isothermal and ideally mixed with respect to the solids. The hydrodynamic profile of the reactor is described by a prescribed solids concentration along the reactor height. Plug flow is assumed for the gases. This may seem a rather simplified structure allowing only little room for adaption. The programming, however, has been done in a way that allows the addition of details in later stages (object orientated code in C++). For the moment, the simple structure avoids unknown parameters that would require rough estimation from widely scattered literature data. Various combinations of parameter values would lead to similar simulation output. Moreover, the present model allows extremely fast calculations. The calculation time for modeling a single reactor (air reactor and fuel reactor respectively) is in the range of a few seconds, whereas the whole reactor system is modeled in $1 - 2\,\mathrm{min}$ with a simple $1.8\,\mathrm{GHz}$ processor. The complete CLC reactor system is modeled by combining two reactor models (i.e. air and fuel reactors).

5.2.2 Reaction model

The smallest unit possible in the reaction model is a single reactor cell. Each cell has a certain volume with an inlet and outlet for a gas and a solid stream. The composition at the cell inlet is known from the previous cell, the outlet composition is the unknown vector. Energy and solids balances can be performed in each cell but for certain reactor

	reaction	reactor	reference
1	$CH_4 + 4NiO \rightleftharpoons CO_2 + 2H_2O + 4Ni$	FR	[46]
2	$CO + NiO \rightleftharpoons CO_2 + Ni$	FR	[46]
3	$H_2 + NiO \rightleftharpoons H_2O + Ni$	FR	[46]
4	$CH_4 + H_2O \rightleftharpoons CO + 3H_2$	FR	[83]
5	$CO + H_2O \rightleftharpoons H_2 + CO_2$	FR	[84]
6	$2Ni + O_2 \rightleftharpoons 2NiO$	AR	[46]

Table 5.1: Oxygen carrier reactions considered in the reactor model

types (such as fluidized beds) this option is disabled. Owing to the assumed perfect mixing character of the fluidized bed reactor mentioned above, solids and energy balances are solved globally.

For each cell, a system of equations has to be solved. For each variable (molar fraction of reacting gaseous and solid components), there exists a reaction rate equation. Some of the reaction rate equations can be replaced by element balances of the occurring elements. The model can generally handle homogeneous gas phase reactions, gas phase reactions catalyzed by the solids and heterogeneous reactions involving the solids as reactants. N^{th} order and hyperbolic kinetic formulations are applied for intrinsic kinetics. The heterogeneous reactions are dependent on the surface of reacting solids present in the control volume according to the shrinking core model.

The reactions actually considered in the present study are summarized in Table 5.1. Each reaction can individually be implemented in the sub-model. In cases where the oxygen carrier is not fully oxidized, metallic Ni is present in the FR. In this case, the steam reforming reaction of CH_4 (Equation (4) in Table 5.1) is catalyzed and should have some impact. Kinetic parameters for this reaction in the presence of the oxygen carrier, however, are not available and thus the parameters determined by Xu and Froment [83] for a Ni-based catalyst are used. This will certainly have some impact on the accuracy of the modeling results but owing to the lack of precise data from literature, this inaccuracy has to be accepted.

The differential equations are solved in an implicit way by iterating the concentrations at the exit of the cell until the gradients (calculated using the estimated exit concentrations) fit in order to arrive from the given input concentrations to the solution for the exit concentration. The iteration is done using a multidimensional Newton method with analytical evaluation of the Jacobian matrix. This way, asymptotic stability is achieved and the size of the cells is not limited by the stability of the solver. The cell size can be chosen as large as possible with respect to linearization errors.

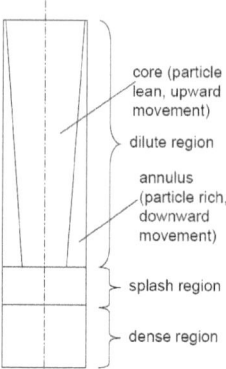

Figure 5.1: Typical structure of the fluid dynamic description of turbulent and fast fluidized bed reactors

5.2.3 Fluid dynamic model

The fluid dynamic description of the fluidized bed reactor is based on the idea of a dense bottom bed followed by a splash region and a lean zone showing a core-annulus type flow structure [77, 85–89] (see Figure 5.1). The core-annulus structure in the upper part of the riser is characterized by high gas velocities in the particle lean core of the riser tube and a particle dense annulus region at the walls. In the core, the particles move upwards whereas in the annulus the particles may either move up or downwards depending on particle size and superficial gas velocity. The local gas velocity in the annulus is low and governed by the particle flow.

Since most gas moves in the core and the solids hold-up is concentrated in the annulus, the core-annulus flow structure of fast fluidized bed risers may limit the gas-solid reaction rate. Parameters have been proposed to account for the exchange of gas and solids between core and annulus along the reactors height [90–92]. Beside these parameters, such a model requires the determination or estimation of several further parameters that all significantly affect the progress of gas-solid reactions in the reactor. As described by Kaushal et al. [87, 88] these include (amongst others)

- the bubble-emulsion mass exchange inside the dense region,

- the bubble-emulsion interfacial area,

- the height of the dense region and
- the solids concentration in the core.

For the evaluation of most of these values, a number of different relations have been published. The derived values, however, scatter strongly in some cases. Additionally, the chemical kinetics and local mass transfer phenomena on the solid particles are to be determined either from suitable literature data or by direct comparison to experimental results with a certain oxygen carrier material. Some of the preceding reactions in the fuel reactor, such as the partial oxidation of CH_4,

$$CH_4 + NiO \rightleftharpoons CO + 2H_2 + Ni \qquad (5.1)$$

have to be neglected owing to the lack of kinetic models available in literature. For other important reactions, such as the steam reforming reaction (see Equation (4) in Table 5.1), kinetic models from different catalysts have to be applied.

It is very uncertain that such complex models can be successfully applied to the 120 kW chemical looping combustor at Vienna University of Technology. Beside the problem of finding reliable model parameter data in literature, it is highly doubtful that the model can be validated with the existing measuring equipment at the pilot rig. Therefore, the model used aims at a most simple description of phase flow behavior possible in order to reduce the number of uncertain parameters to an absolute minimum.

The main point where macroscopic fluid dynamics affect the progress of chemical reactions is where the exposure of the gas, passing the reactor, to the solids is quantified. This is qualitatively valid in each of the regions shown in Figure 5.1. There are, of course, other potentially relevant phenomena like axial back mixing in the gas phase that may reduce conversion by reducing chemical driving forces. A detailed view, however, would require tracer measurements for validation and goes beyond the scope of the present study. The solids can be assumed to be rather well mixed with respect to composition, providing that the chemical conversion of the solids is slow in comparison to particle displacement.

Therefore, it is assumed that the macroscopic fluid dynamics of the fluidized bed reactor affect the gas conversion in proportion to a certain fraction of solids actually contacted to the gas. For a given solids hold-up in the reactors and a prescribed axial solids concentration profile, the model parameter $\phi_{s,core}$ defines the fraction of solids exposed to the gas passing in plug flow. Many uncertain parameters are combined in this single parameter, which is further assumed to be constant over the whole reactor height (see Figure 5.2). The solids composition is the same throughout the reactor and

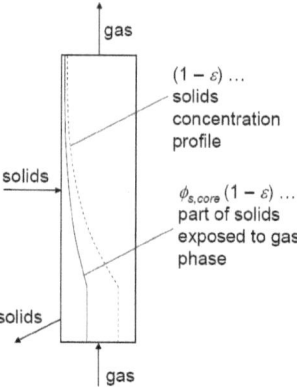

Figure 5.2: Simplified core-annulus model with only one parameter ($\phi_{s,core}$)

changes with reaction progress (considering the entire solids hold-up). Of course, this very simple model cannot explain all effects along the height axis of a circulating fluidized bed in detail but it will at least allow the determination of $\phi_{s,core}$ for the case of the 120 kW test rig. Further, as mentioned previously, there are many uncertain aspects in the applied reactions and their kinetic parameters. Thus, the gain in quality by applying a more complex fluid dynamic description of the CFB, may be questioned. The effect of axial solids concentration profiles (e.g. described by Schlichthaerle and Werther [93]) is neglected in this study.

The solids distribution along the reactor height is assumed to follow a typical dense zone/lean zone structure. In the lean zone, an exponential decay of solids hold-up is assumed. Therefore, the solids concentration along the reactor height is defined by the following equations:

$$\epsilon(h) = \epsilon_{DZ} \quad \text{in the dense zone } (h \leq h_{DZ}) \tag{5.2}$$

$$(1 - \epsilon(h)) = (1 - \epsilon_{DZ}) \cdot e^{-\alpha(h - h_{DZ})} \quad \text{in the dilute zone} \tag{5.3}$$

These profiles can be determined at the 120 kW pilot rig with the aid of pressure measurements along the reactor height.

5.2.4 Energy balance

The temperature distribution in the system is dependent on the heat production (conversion of the fuel), the inlet temperatures of air and fuel, the solids circulation rate and the cooling power in each reactor. The energy balance of each reactor can be written as

$$(\sum_i \dot{m}_i \cdot h_i)_{in} - (\sum_j \dot{m}_j \cdot h_j)_{out} - \dot{Q} = 0 \qquad (5.4)$$

with the mass flow $\dot{m}$, the enthalpy of formation h of each inlet and outlet stream and the cooling duty $\dot{Q}$. Note that in equation (5.4) both, gas and solid streams, have to be considered.

While the composition of the entering gas streams is given, the composition of the exiting gas streams and all solid streams are determined by the chemical reactions. The solids circulation rate is prescribed. Assuming constant temperature in each reactor, one obtains two equations for the four unknown variables T_{AR}, $\dot{Q}_{AR}$, T_{FR} and $\dot{Q}_{FR}$. In the present study the fuel reactor is regarded as adiabatic ($\dot{Q}_{FR} = 0$) with a set temperature. From these definitions, the air reactor temperature and cooling duty can be calculated for stable operation.

5.3. Modeling results

The introduced model is applied to study the influence of different operating parameters on the performance of the 120 kW CLC pilot rig at Vienna University of Technology. Owing to the lack of kinetic parameters of the oxygen carrier used in the pilot rig, a slightly different oxygen carrier is used (Ni40Al-FG [46]). The active NiO content and oxygen transport capacity of the oxygen carriers, however, are very similar. The main oxygen carrier characteristics are shown in Table 5.2.

Table 5.3 summarizes the applied solids distribution and the main operating parameters of the air and fuel reactors. In this work, the influence of the model parameter $\phi_{s,core}$ on the solids conversion in the fuel reactor ($X_{S,FR}$) is presented only. Further results which include the determination of gas conversions, different parameter variations and a sensitivity analysis are presented in Paper VI.

The investigation of the solids conversion in the reactor system very impressively shows that, even though a very simple hydrodynamic model is applied, the results reflect most actual conditions in the CLC pilot rig. Figure 5.3 illustrates the solids conversion in the fuel reactor. Assuming similar values of $\phi_{s,core}$ in the air and fuel reactors and incomplete conversion ($\phi_{s,core,AR,FR} < 0.25$), the OC particles leave the fuel reactor

parameter	unit	value
designation		Ni40Al-FG
total NiO content (in raw material)	wt%	60
active NiO content	wt%	40
support material		$NiAl_2O_4$
oxygen transport capacity $R_{0,OC}$	kg/kg	0.084
apparent density	kg/m^3	3446
particle size	mm	0.2
specific BET surface area	m^2/g	0.8
porosity	–	0.36

Table 5.2: Main properties of the oxygen carrier applied for modeling [46]

with a solids conversion in the range of $0.20 - 0.65$. With the given solids circulation rate this corresponds to a solids conversion of $0.30 - 0.75$ in the air reactor. In the CLC pilot rig, $X_{S,FR}$ is usually determined in the range of $0.1 - 0.3$ when OC-A is applied.[1] Therefore, it is expected that the model parameter $\phi_{s,core}$ in the air and fuel reactors is in the range of $0.05 - 0.20$. Of course, this is just a rough estimate and does not replace proper validation of the model.[2]

param.	description	unit	AR	FR
m_S	solids hold-up	kg	17.6	19.4
h_{DZ}	dense zone height	m	0.5	0.5
ε_{DZ}	dense zone void fraction	–	0.75	0.75
α	solids conc. decay factor in lean zone	m^{-1}	1.4	1.4
$\triangle p$	pressure drop	kPa	9.8	9.7
T	reactor temperature	°C	E-bal.*	850
P_{th}	fuel power	kW	120	
λ	air/fuel ratio	–	1.2	
S/C	steam/methane ratio	mol/mol	0.1	
G_S	global solids circulation rate	kgm^{-2}s^{-1}	50.0	—

* the AR temperature is calculated from the energy balance of the reactor system

Table 5.3: Reference case for modeling of the CLC pilot rig

---

[1] OC-A is more similar to the oxygen carrier used in this study.
[2] Validation, however, is not included in this work.

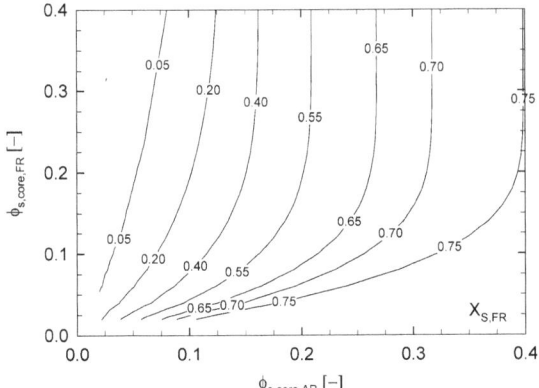

Figure 5.3: Solids conversion in the fuel reactor as a function of the model parameters $\phi_{s,core,AR}$ and $\phi_{s,core,FR}$

5.4. General aspects on modeling of continuous looping systems

For predictive modeling and simulation of chemical looping processes, the kinetic parameters of gas-solid reactions have to be determined. Different authors have applied the shrinking core model (SCM) with the reaction controlled by the chemical reaction in the grain for spherical particles (e.g. Abad et al. [45], Grasa et al. [94]) and plate-like particles (e.g. Garcia-Labiano et al. [47], Zafar et al. [48]). When applying such models for the modeling of continuous looping operation, however, care must be taken in the formulation of reaction rate equations.

Garcia-Labiano et al. [47] have already pointed out that kinetic models from batch experiments have to be adjusted for use in modeling of continuous looping operation. In their work, which focuses on Cu-based particles, the reaction rates in the air and fuel reactors are formulated in a way that the influence of a prescribed particle residence time distribution, the solids conversion of the entering solids to a reactor and the difference in solids conversion between air and fuel reactors are considered. Abad et al. [45, 46] have introduced the so-called characteristic reactivity which also considers these influences. In the following, different macroscopic phenomena are discussed that have to be considered, for the precise formulation of reaction rates for continuous looping operation.

In the SCM, it is assumed that each particle consists of a number of spherical grains. Reactions occur at the outer surface of the grain, reducing its diameter and outer surface

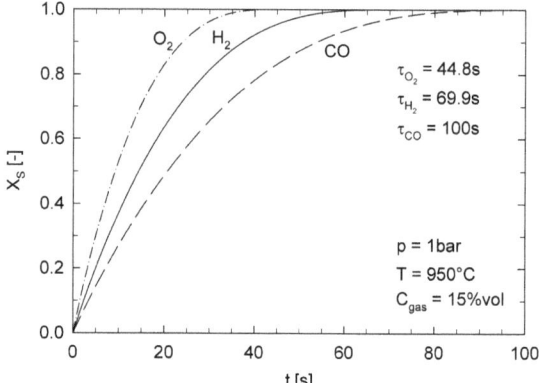

Figure 5.4: Solids conversion of an oxygen carrier for CLC as a function of elapsed time. $X_S = 1.0$ corresponds to the fully converted particle with respect to the used gas (i.e. fully oxidized for O_2 and fully reduced for H_2 and CO respectively). Kinetics data have been taken from [45].

area as they proceed. Since the reaction rate is only controlled by the outer surface of the grain (diffusion resistance of gaseous reactants and products through the gas film and "ash" is neglected), it will diminish as the grain shrinks [95].

From the obtained experimental data, X_S vs. time diagrams can be plotted and fitted with an adequate model (such as the SCM). An example of such a graph is plotted in Figure 5.4 for the reduction and oxidation of a Ni-based oxygen carrier particle for CLC according to Abad et al. [45]. Depending on the applied gas, the time for full conversion varies. The curve progression, however, is identical for all gases. In this case, the slope corresponds to the SCM for a spherical grain with the reaction controlled by the chemical reaction in the grain.

In the case of continuous operation, the particles are only partially converted until they pass back to the second reactor. Therefore, only a relatively small part of the cyclic capacity is actually used. The particles are reduced to $X_{S,FR}$ in the fuel reactor and oxidized to $X_{S,AR}$ in the air reactor. The difference between the two states, ΔX_S, directly results from the solids circulation rate of the reactor system.

If a reactor model of such a continuous unit uses data determined in batch experiments, the gas-solid reaction rates in the reactors should, according to the model, depend on the actual degree of particle conversion X_S. Figure 5.5 illustrates an oxidation-reduction

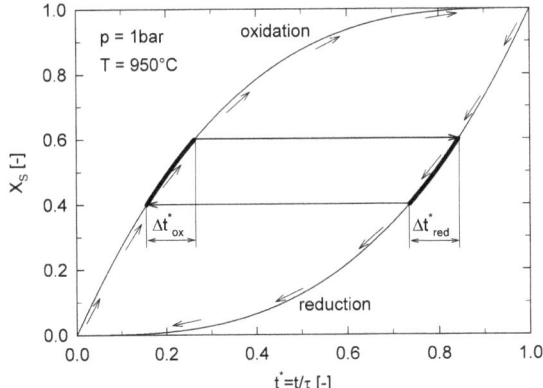

Figure 5.5: Loop of a single particle in a solids conversion vs. time diagram. The cycle from $X_S = 0.0$ to 1.0 and back represents a batch experiment cycle. A chemical looping cycle is characterized by much smaller solids conversion differences.

cycle and the reaction rates (dX_S/dt) used if kinetics data from batch experiments are implemented directly. When visualizing the macroscopic changes during batch experiments and continuous looping experiments, however, one will discover fundamental differences between these approaches. In batch tests, a loop starts with a fully unconverted particle. At first the gas will react with the sites that are most easily accessible and will advance to sites more difficult to access. The reaction of the latter will naturally be slower than the former. The conversion of the particle will eventually come to a stop when the particle is fully converted.

In continuous looping operation, however, the situation looks somewhat different. The particle enters the fuel reactor with a certain degree of solids conversion $X_{S,AR}$. Again, the reduction of the particle will start at the sites that are easiest to access. These are, however, very probably the same as in the case where the particle would have been fully oxidized before, i.e. $X_{S,AR} = 1.0$ (even though $X_{S,AR}$ might be much smaller). The particle is reduced to $X_{S,FR}$ and transported back to the air reactor where it is again oxidized. Once more, conversion starts at the easily accessible sites which are, again, the same as in the case of a fully reduced particle. Thus, the continuous process will, very probably, operate with the outer grain sites only.

Given these macroscopic effects, a single cycle of a particle in continuous looping op-

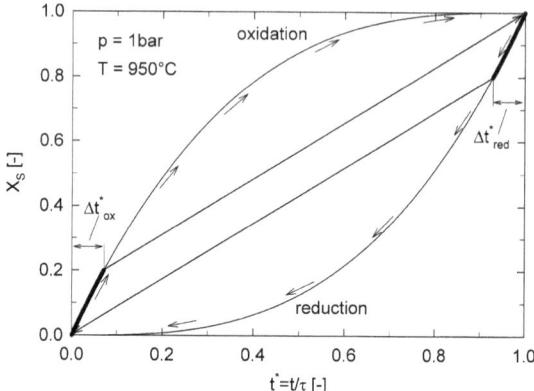

Figure 5.6: Loop of a single particle in a solids conversion vs. time diagram with the consideration of the macroscopic particle behavior. The time for the conversion of a certain $\triangle X_S$ ($\triangle t^*_{ox}$ and $\triangle t^*_{red}$, respectively) decreases since the reaction rate is increased in this case.

eration must look rather more like that in Figure 5.6 in terms of reaction rates than previously suggested in Figure 5.5. One can clearly see that the time to convert the particle by $\triangle X_S$ is much smaller than previously expected. This derives from the increased reaction rate in this case. At this point, one starts to realize that one single parameter, such as the solids conversion, might not be sufficient for the description of the state of the particle.[3]

The particle cycles mentioned in this study will probably not have much influence on the evaluation of the performance of a chemical looping reactor, where the main interest is the gas conversion. When it comes to detailed modeling of such processes, however, care must be taken in the formulation of reaction rates of particles as a function of solids conversion. Further, these effects have some impact on the required solids inventory and have to be considered in the design phase of chemical looping installations.

---

[3] This is especially true for continuous looping operation.

6

CONCLUSIONS AND OUTLOOK

Chemical looping combustion (CLC) has been identified as a high-potential CO_2 capture technology. CLC reactor systems are composed of two interconnected fluidized beds with an oxygen carrier as bed material. Excellent gas-solids contact in the reaction zones to ensure sufficient gas conversion and high solids circulation are essential for successful operation of a CLC plant. The dual circulating fluidized bed (DCFB) reactor system meets the demands very well and has a high potential for scale-up.

A DCFB reactor system for chemical looping combustion with a nominal fuel power of 120 kW has been built at Vienna University of Technology. This unit can be operated with CH_4, CO, H_2 and C_3H_8 as fuel. In the first year of operation, this pilot rig has shown

- high qualification of the reactor concept for chemical looping combustion (high solids circulation rate, sufficient gas-solids contact),

- full conversion of H_2 and CO with a Ni-based oxygen carrier,

- very high CH_4 conversion (close to 99 %) and CO_2 yield (close to 95 %) with Ni-based oxygen carriers,

- ilmenite as a potential oxygen carrier for the combustion of syngases and

- no carbon formation in any of the tested conditions.

Quite surprisingly, a very low solids conversion of the particles has been determined throughout the reactor system. This indicates that in the built unit the air reactor is the limiting part in the system.[1] A modeling tool, also developed within this project, has predicted these conditions very well.

Even greater improvement of the combustion efficiency is expected for higher air and fuel reactors. Unfortunately, the reactor height is limited owing to the surrounding laboratory (i.e. $h_{AR} = 4.1\,\text{m}$ and $h_{FR} = 3.0\,\text{m}$). Circulating fluidized beds, however,

---

[1] This does not mean, however, that a larger air reactor will improve the combustion efficiency.

cannot be scaled in height. Therefore, the pilot rig height should be the same as in industrial applications.

Despite the recent development and the results obtained, there are still a number of aspects that should be investigated in the near future. These include

- further tests with ilmenite and other cheap natural minerals,
- the investigation of mixed oxides (e.g. ilmenite in different mixtures with Ni-based particles),
- chemical looping reforming tests,
- the investigation of carbon formation at very low air/fuel ratios and temperatures and
- long-term experiments to investigate possible changes in particle properties.

The next step for chemical looping combustion of gaseous fuel is the demonstration of the technology on a larger scale (approx. 10 MW). Some niche markets, such as the combustion of offgases from petrochemistry and process steam production, offer particularly good opportunities for gas-fired CLC. Certainly, there is still much work to be done and experiments to be performed to reach this next step. Nevertheless, the results obtained in the 120 kW CLC pilot rig are already highly promising and clearly point to the large potential of this novel technology.

7

LIST OF PUBLICATIONS

This thesis is based on the work published in the following articles:

I. Kolbitsch P., Bolhar-Nordenkampf J., Pröll T., Hofbauer H. Design of a chemical looping combustor using a dual circulating fluidized bed (DCFB) reactor system. *Chemical Engineering and Technology* 32(3), **2009**, 398–403.

II. Pröll T., Kolbitsch P., Bolhar-Nordenkampf J., Hofbauer H. A novel dual circulating fluidized bed (DCFB) system for chemical looping processes. *accepted for publication in AIChE Journal* **2009**.

III. Kolbitsch P., Pröll T., Bolhar-Nordenkampf J., Hofbauer H. Operating experience with chemical looping combustion in a 120kW dual circulating fluidized bed (DCFB) unit. *Energy Procedia* 1(1), **2009**, 1465–1472.

IV. Kolbitsch P., Pröll T., Bolhar-Nordenkampf J., Hofbauer H. Characterization of chemical looping pilot plant performance via experimental determination of solids conversion. *Energy & Fuels* 23, **2009**, 1450–1455.

V. Kolbitsch P., Bolhar-Nordenkampf J., Pröll T., Hofbauer H. Comparison of two Ni-based oxygen carriers for chemical looping combustion of natural gas in 140 kW continuous looping operation. *Industrial & Engineering Chemistry Research*, **2009**, DOI:10.1021/ie900123v.

VI. Kolbitsch P., Pröll T., Hofbauer H. Modeling of a 120kW chemical looping combustion reactor system using a Ni-based oxygen carrier. *Chemical Engineering Science* 64(1), **2009**, 99–108.

Related articles:

i. Kolbitsch P., Pröll T., Hofbauer H. On the applicability of data from batch experiments for the prediction of reaction rates in continuous chemical looping systems. *submitted to Industrial & Engineering Chemistry Research*, **2009**.

ii. Pröll T., Mayer K., Bolhar-Nordenkampf J., Kolbitsch P., Mattisson T., Lyngfelt A., Hofbauer H. Natural minerals as oxygen carriers for chemical looping combustion in a dual circulating fluidized bed system. *Energy Procedia* 1(1), **2009**, 27–34.

iii. Bolhar-Nordenkampf J., Pröll T., Kolbitsch P., Hofbauer H. Performance of a NiO-based oxygen carrier for chemical looping combustion and reforming in a 120kW unit. *Energy Procedia* 1(1), **2009**, 19–25.

iv. Pröll T., Rupanovits K., Kolbitsch P., Bolhar-Nordenkampf J., Hofbauer H. Cold flow model study on a dual circulating fluidized bed (DCFB) system for chemical looping processes. *Chemical Engineering and Technology* 32(3), **2009**, 418–424.

v. Bolhar-Nordenkampf J., Pröll T., Kolbitsch P., Hofbauer H. Comprehensive modeling tool for chemical looping based processes. *Chemical Engineering and Technology* 32(3), **2009**, 410–417.

vi. Pröll T., Kolbitsch P., Bolhar-Nordenkampf J., Hofbauer H. A dual circulating fluidized bed (DCFB) system for chemical looping combustion. In *Proceedings of the 2008 AIChE Annual Meeting*, **2008**, Philadelphia, USA.

vii. Pröll T., Bolhar-Nordenkampf J., Kolbitsch P., Marx K., Hofbauer H. Pure hydrogen and pure carbon dioxide from gaseous hydrocarbons by chemical looping reforming. In *Proceedings of the 2009 AIChE Spring National Meeting*, **2009**, Tampa, USA.

viii. Bolhar-Nordenkampf J., Pröll T., Kolbitsch P., Hofbauer H. Chemical looping combustion for power generation - concept study for a 10MWth demonstration plant. *submitted to Combustion and Flame* **2009**.

ix. Bolhar-Nordenkampf J., Pröll T., Kolbitsch P., Hofbauer H. Chemical looping reforming for syngas generation from natural gas - Based on results from a 120kW fuel power installation. *accepted for publication in Syngas: Production Methods, Post Treatment and Economics*, Nova Science Publishers, Inc., **2009**.

x. Bolhar-Nordenkampf J., Pröll T., Kolbitsch P., Hofbauer H. Chemical looping autothermal reforming at a 120kW pilot rig. In *Proceedings of 20th International Conference on Fluidized Bed Combustion*, **2009**, Xian, China.

8

NOTATION

8.1. Abbreviations

AR	air reactor
ASU	air separation unit
BFB	bubbling fluidized bed
CCS	carbon capture and storage
CFB	circulating fluidized bed
CFBC	circulating fluidized bed combustion
CLC	chemical looping combustion
CLR	chemical looping reforming
CLR(a)	chemical looping autothermal reforming
CLR(s)	chemical looping steam reforming
DCFB	dual circulating fluidized bed
DZ	dense zone of the fluidized bed
EOR	enhanced oil recovery
FCC	fluid catalytic cracking
FR	fuel reactor
GHG	greenhouse gas
GWP	global warming potential
IEA	International Energy Agency
IGCC	Integrated Gasification Combined Cycle
ILS	internal loop seal
IPCC	International Panel on Climate Change
LGP	liquefied petroleum gas
LHV	lower heating value
LLS	lower loop seal
ng	natural gas
OECD	Organization for Economic Co-operation and Development
OC	oxygen carrier
PSA	pressure swing adsorption
S/C	steam to carbon (methane) molar ratio
SCM	shrinking core model

SOFC	solid oxide fuel cell
SRES	Special Report on Emission Scenarios
TGA	thermo-gravimetric analyzer
ULS	upper loop seal
UNFCCC	United Nations Framework Convention on Climate Change
WGS	water gas shift (reaction)

8.2. Symbols

A	cross section	m²
C_W	drag coefficient	(−)
d	diameter	m
f_p	friction factor	(−)
g	standard gravity (9.80665)	m/s²
G_S	net solids flux	kg/m²s
h	height	m
h_i	enthalpy of formation of substance i	J/kg
L	length	m
M_i	molar mass of the substance i	g/mol
m_i	mass of substance i	kg
$\dot{m}_i$	mass flow of substance i	kg/s
$\dot{m}_i$	molar flow of substance i	mol/s
O_{min}	O_2 demand for fuel oxidation	kg/kg
p	pressure	Pa
P	power	W
P_{th}	fuel power	W
$\dot{Q}$	heat loss	W
R_0	oxygen transport capacity	kg/kg
Re	Reynolds number	(−)
T	temperature	K, °C
U	velocity	m/s
x	volumetric fraction (of gases)	(−)
X	(gas, solids) conversion	(−)
α	solids conc. decay factor in lean zone	1/m
ε	void fraction	(−)
λ	air/fuel ratio	(−)
μ	kinematic viscosity	m²/s
ρ	density	kgm^{-3}

τ	(gas, solids) residence time	s
τ	time for complete combustion	s
$\phi_{s,core}$	model parameter	(−)

8.3. Sub and superscripts

*	dimensionless parameter
DZ	dense zone
gas	valid for gas phase
in	inlet stream
out	outlet stream
ox	oxidized form
p	particle
red	reduced form
S	solid phase

9

REFERENCES

[1] Fourier J. Mémoires sur les températures du globe terrestre et des espaces planétaires. *Memoire de l'Academie royale des sciences de l'Institut de France* VII, **1827**, 570–604.

[2] Tyndall J. On the absorption and radiation of heat by gases and vapours. *Philosophical Magazine* IV, **1861**, 169–194 and 273–285.

[3] Arrhenius S. On the Influence of Carbonic Acid in the Air upon the Temperature of the Ground. *Philosophical Magazine and Journal of Science* 41, **1896**, 237–276.

[4] IPCC. *Climate Change 2007: The Physical Science Basis. Contribution of Working Group I to the Fourth Assessment Report of the Intergovernmental Panel on Climate Change.* Cambridge University Press, Cambridge, United Kingdom and New York, NY, USA, **2007**.

[5] IPCC. *Special report emissions scenarios.* Cambridge University Press, Cambridge, United Kingdom and New York, NY, USA, **2000**.

[6] Joos F., Prentice I.C., Sitch S., Meyer R., Hooss G., Plattner G.K., Gerber S., Hasselmann K. Global warming feedbacks on terrestrial carbon uptake under the Intergovernmental Panel on Climate Change (IPCC) emission scenarios. *Global Biogeochemical Cycles* 15, **2001**, 891–907.

[7] Birol F. *World Energy Outlook 2006.* International Energy Agency (IEA), **2006**.

[8] Working Group III of the Intergovernmental Panel on Climate Change. *IPCC Special Report on Carbon Dioxide Capture and Storage.* Cambridge University Press, Cambridge, United Kingdom and New York, NY, USA, **2005**.

[9] McKee B. *Solutions for the 21st century, Zero Emission Technologies for Fossil Fuels.* IEA Technology status report, **2002**.

[10] U.S. Department of Energy. *Carbon sequestration technology roadmap and program plan.* **2007**.

[11] Lyon R.K., Cole J.A. Unmixed Combustion: An Alternative to Fire. *Combustion & Flame* 121, **2000**, 249–261.

[12] Davison J. Performance and costs of power plants with capture and storage of CO2. *Energy* 32(7), **2007**, 1163–1176.

[13] Ishida M., Jin H. A Novel Chemical-Looping Combustor without NOx Formation. *Industrial & Engineering Chemistry Research* 35, **1996**, 2469–2472.

[14] Ryu H., Jin G., Yi C. Demonstration of inherent CO2 separation and no NOx emission in a 50 kW Chemical-Looping Combustor: Continuous reduction and oxidation experiment. In *7th Conference on Greenhouse Gas Control Technologies (GHGT7)*. Vancouver, Canada, **2004**.

[15] Lyngfelt A., Leckner B., Mattisson T. A fluidized-bed combustion process with inherent CO2 separation; application of chemical-looping combustion. *Chemical Engineering Science* 56(10), **2001**, 3101–3113.

[16] Adanez J., Garcia-Labiano F., de Diego L.F., Gayan P., Celaya J., Abad A. Nickel-Copper Oxygen Carriers To Reach Zero CO and H2 Emissions in Chemical-Looping Combustion. *Industrial & Engineering Chemistry Research* 45(8), **2006**, 2617–2625.

[17] Hossain M.M., de Lasa H.I. Reactivity and stability of Co-Ni/Al2O3 oxygen carrier in multicycle CLC. *AIChE Journal* 53(7), **2007**, 1817–1829.

[18] Johansson M., Mattisson T., Lyngfelt A. Creating a Synergy Effect by Using Mixed Oxides of Iron- and Nickel Oxides in the Combustion of Methane in a Chemical-Looping Combustion Reactor. *Energy & Fuels* 20(6), **2006**, 2399–2407.

[19] Jerndal E., Mattisson T., Lyngfelt A. Thermal analysis of chemical-looping combustion. *Chemical Engineering Research and Design* 84, **2006**, 795–806.

[20] Adanez J., de Diego L.F., Garcia-Labiano F., Gayan P., Abad A., Palacios J.M. Selection of Oxygen Carriers for Chemical-Looping Combustion. *Energy & Fuels* 18(2), **2004**, 371–377.

[21] Johansson M., Mattisson T., Rydn M., Lyngfelt A. Carbon Capture via Chemical-Looping Combustion and Reforming. In *International Seminar on Carbon Sequestration and Climate Change*. Rio de Janeiro, **2006**.

[22] Lewis W.K., Gilliland E.R. Production of pure carbon dioxide. U.S. Patent Office, Number 2,665,972, **1954**.

[23] Knoche K., Richter H. Verbesserung der Reversibilitaet von Verbrennungsprozessen. *Brennstoff-Waerme-Kraft* 20(5), **1968**, 205–211.

[24] Richter H.J., Knoche K.F. Reversibility of combustion processes. *ACS Symposium Series* 235, **1983**, 71–85.

[25] Ishida M., Zheng D., Akehata T. Evaluation of a chemical-looping-combustion power-generation system by graphic exergy analysis. *Energy* 12(2), **1987**, 147–154.

[26] Ishida M., Jin H. A new advanced power-generation system using chemical-looping combustion. *Energy* 19(4), **1994**, 415–422.

[27] Lyngfelt A., Johansson M., Mattisson T. Chemical-looping combustion - Status and development. In *9th International Conference on Circulating Fluidized Beds*. **2008**, 39–53.

[28] Lyngfelt A., Thunman H. Construction and 100h of operational experience of a 10-kW chemical looping combustor. In *Carbon Dioxide Capture for Storage in Deep Geologic Formations*, chap. 31. Elsevier Science, Amsterdam, **2005**, 625–645.

[29] Berguerand N., Lyngfelt A. Design and operation of a 10kWth chemical-looping combustor for solid fuels - Testing with South African coal. *Fuel* 87, **2008**, 2713–2726.

[30] Ryden M., Lyngfelt A., Mattisson T. Synthesis gas generation by chemical-looping reforming in a continuously operating laboratory reactor. *Fuel* 85, **2006**, 1631–1641.

[31] de Diego L.F., Garcia-Labiano F., Gayan P., Celaya J., Palacios J.M., Adanez J. Operation of a 10kWth chemical-looping combustor during 200h with a CuO-Al2O3 oxygen carrier. *Fuel* 86, **2007**, 1036–1045.

[32] Son S.R., Kim S.D. Chemical-Looping Combustion with NiO and Fe2O3 in a Thermobalance and Circulating Fluidized Bed Reactor with Double Loops. *Industrial & Engineering Chemistry Research* 45(8), **2006**, 2689–2696.

[33] Cho P., Mattisson T., Lyngfelt A. Comparison of iron-, nickel-, copper- and manganese-based oxygen carriers for chemical-looping combustion. *Fuel* 83(9), **2004**, 1215–1225.

[34] Cho P., Mattisson T., Lyngfelt A. Carbon Formation on Nickel and Iron Oxide-Containing Oxygen Carriers for Chemical-Looping Combustion. *Industrial & Engineering Chemistry Research* 44(4), **2005**, 668–676.

[35] Cho P., Mattisson T., Lyngfelt A. Defluidization Conditions for a Fluidized Bed of Iron Oxide-, Nickel Oxide-, and Manganese Oxide-Containing Oxygen Carriers for Chemical-Looping Combustion. *Industrial & Engineering Chemistry Research* 45, **2006**, 968–977.

[36] Corbella B.M., De Diego L., Garcia F., Adanez J., Palacios J.M. Characterization and Performance in a Multicycle Test in a Fixed-Bed Reactor of Silica-Supported Copper Oxide as Oxygen Carrier for Chemical-Looping Combustion of Methane. *Energy & Fuels* 20, **2006**, 148–154.

[37] de Diego L.F., Garcia-Labiano F., Adanez J., Gayan P., Abad A., Corbella B., Palacios J. Development of Cu-based oxygen carriers for chemical-looping combustion. *Fuel* 83, **2004**, 1749–1757.

[38] de Diego L.F., Gayan P., Garcia-Labiano F., Celaya J., Abad A., Adanez J. Impregnated CuO/Al2O3 Oxygen Carriers for Chemical-Looping Combustion: Avoiding Fluidized Bed Agglomeration. *Energy & Fuels* 19(5), **2005**, 1850–1856.

[39] Ishida M., Takeshita K., Suuki K., Ohba T. Application of Fe2O3-Al2O3 Composite Particles as Solid Looping Material of the Chemical-Loop Combustor. *Energy & Fuels* 19, **2005**, 2514–2518.

[40] Johansson M., Mattisson T., Lyngfelt A. Investigation of Fe2O3 with MgAl2O4 for Chemical-Looping Combustion. *Industrial & Engineering Chemistry Research* 43(22), **2004**, 6978–6987.

[41] Abad A., Mattisson T., Lyngfelt A., Ryden M. Chemical-looping combustion in a 300W continuously operating reactor system using a manganese-based oxygen carrier. *Fuel* 85(9), **2006**, 1174–1185.

[42] Abad A., Mattisson T., Lyngfelt A., Johansson M. The use of iron oxide as oxygen carrier in a chemical-looping reactor. *Fuel* 86(7-8), **2007**, 1021–1035.

[43] Johansson M., Mattisson T., Lyngfelt A. Use of NiO/NiAl2O4 Particles in a 10 kW Chemical-Looping Combustor. *Industrial & Engineering Chemistry Research* 45(17), **2006**, 5911–5919.

[44] Linderholm C., Abad A., Mattisson T., Lyngfelt A. 160h of chemical-looping combustion in a 10kW reactor system with a NiO-based oxygen carrier. *International Journal of Greenhouse Gas Control* 2(4), **2008**, 520–530. TCCS-4: The 4th Trondheim Conference on CO2 Capture, Transport and Storage.

[45] Abad A., Garcia-Labiano F., de Diego L.F., Gayan P., Adanez J. Reduction Kinetics of Cu-, Ni-, and Fe-Based Oxygen Carriers Using Syngas (CO + H2) for Chemical-Looping Combustion. *Energy & Fuels* 21(4), **2007**, 1843–1853.

[46] Abad A., Adanez J., Garcia-Labiano F., de Diego L.F., Gayan P., Celaya J. Mapping of the range of operational conditions for Cu-, Fe-, and Ni-based oxygen carriers in chemical-looping combustion. *Chemical Engineering Science* 62(1-2), **2007**, 533–549.

[47] Garcia-Labiano F., De Diego L.F., Adanez J., Abad A., Gayan P. Reduction and Oxidation Kinetics of a Copper-Based Oxygen Carrier Prepared by Impregnation for Chemical-Looping Combustion. *Industrial & Engineering Chemistry Research* 43(26), **2004**, 8168–8177.

[48] Zafar Q., Abad A., Mattisson T., Gevert B., Strand M. Reduction and oxidation kinetics of Mn3O4/Mg-ZrO2 oxygen carrier particles for chemical-looping combustion. *Chemical Engineering Science* 62(2), **2007**, 6556–6567.

[49] Zafar Q., Abad A., Mattisson T., Gevert B. Reaction Kinetics of Freeze-Granulated NiO/MgAl2O4 Oxygen Carrier Particles for Chemical-Looping Combustion. *Energy & Fuels* 21(2), **2007**, 610–618.

[50] Garcia-Labiano F., Adanez J., de Diego L.F., Gayan P., Abad A. Effect of Pressure on the Behavior of Copper-, Iron-, and Nickel-Based Oxygen Carriers for Chemical-Looping Combustion. *Energy & Fuels* 20(1), **2006**, 26–33.

[51] Mattisson T., Lyngfelt A. Application of chemical-looping combustion with capture of CO2. In *2nd Nordic Minisymposium on Carbon Dioxide Capture and Storage*. Göteborg, Sweden, **2001**.

[52] Ryden M., Lyngfelt A., Mattisson T. Chemical-Looping Combustion and Chemical-Looping Reforming in a Circulating Fluidized-Bed Reactor Using Ni-Based Oxygen Carriers. *Energy & Fuels* 22, **2008**, 2585–2597.

[53] Ryden M., Lyngfelt A. Hydrogen and power production with integrated carbon dioxide capture by chemical-looping reforming. In *7th Conference on Greenhouse Gas Control Technologies*. Vancouver, Canada, **2004**.

[54] Ryden M., Lyngfelt A. Using steam reforming to produce hydrogen with carbon dioxide capture by chemical-looping combustion. *International Journal of Hydrogen Energy* 31, **2006**, 1271–1283.

[55] Berguerand N., Lyngfelt A. The use of petroleum coke as fuel in a 10kWth chemical-looping combustor. *International Journal of Greenhouse Gas Control* 2, **2008**, 169–179.

[56] Grace J.R. High-velocity fluidized bed reactors. *Chemical Engineering Science* 45(8), **1990**, 1953–1966.

[57] Hofbauer H., Stoiber H., Veronik G. Gasification of Organic Material in a Novel Fluidized Bed System. In *1st SCEJ Symposium on Fluidization*. Tokyo, Japan, **1995**, 291–299.

[58] Abanades J.C., Anthony E.J., Lu D.Y., Salvador C., Alvarez D. Capture of CO2 from combustion gases in a fluidized bed of CaO. *AIChE Journal* 50(7), **2004**, 1614–1622.

[59] Pfeifer C., Soukup G., Kreuzeder A., Cuadrat A., Hofbauer H. In-situ CO2-capture in a dual fluidized bed biomass steam gasifier: Bed material and fuel variation. In *9th International Conference on Circulating Fluidized Beds*. Hamburg, Germany, **2008**.

[60] Johnsen K., Ryu H., Grace J., Lim C. Sorption-enhanced steam reforming of methane in a fluidized bed reactor with dolomite as CO2-acceptor. *Chemical Engineering Science* 61(4), **2006**, 1195–1202.

[61] Paisley M., Farris M., Black J., Irving J., R.P. O. Preliminary operating results from the Battelle/FERCO gasification demonstration plant in Burlington, Vermont, USA. In *1st World Conference on Biomass for Energy and Industry*. Sevilla, Spain, **2000**.

[62] Andrus H. Chemical looping combustion - R&D efforts by Alstom. In *IEA GHG 2nd Workshop of the International Oxy-Combustion Research Network*. Windsor, USA, **2007**.

[63] Salamov A. Circulating Fluidized Bed Boilers Abroad. *Thermal Engineering* 54(6), **2007**, 501–505.

[64] Pröll T., Rupanovits K., Kolbitsch P., Bolhar-Nordenkampf J., Hofbauer H. Cold flow model study on a dual circulating fluidized bed (DCFB) system for chemical looping processes. *Chemical Engineering and Technology* 32(3), **2009**, 418–424.

[65] Grace J., Bi H. Introduction to circulating fluidized beds. In *Circulating Fluidized Beds*, edited by Grace J., Avidan A., Knowlton T., chap. Introduction to circulating fluidized beds. Chapman & Hall, New York, **1997**, first ed., 1–20.

[66] Kunii D., Levenspiel O. *Fluidization Engineering*. Butterworth-Heinemann series in chemical engineering. Butterworth-Heinemann, Washington, **1991**, second ed.

[67] Geldart D. Types of gas fluidization. *Powder Technology* 7(5), **1973**, 285–292.

[68] Grace J.R. Contacting modes and behavior classification of gas-solid and other two-phase suspensions. *Canadian Journal of Chemical Engineering* 64, **1986**, 353–363.

[69] Stewart P.S.B., Davidson J.F. Slug flow in fluidised beds. *Powder Technology* 1(2), **1967**, 61–80.

[70] Bi H.T., Grace J.R. Flow regime diagrams for gas-solid fluidization and upward transport. *International Journal of Multiphase Flow* 21(6), **1995**, 1229–1236.

[71] Bi H.T., Grace J.R., Zhu J. Regime transitions affecting gas-solids suspensions and fluidized beds. *Chemical Engineering Research and Design* 73, **1995**, 154–161.

[72] Hugi E. *Auslegung hochbeladener Zyklonabscheider für zirkulierende Gas/Feststoff-Wirbelschicht-Reaktorsysteme*. Tech. Rep. Reihe 3, Nr. 502, VDI-Fortschrittsberichte, Düsseldorf, **1997**.

[73] Bolhar-Nordenkampf J., Pröll T., Kolbitsch P., Hofbauer H. Comprehensive modeling tool for chemical looping based processes. *Chemical Engineering and Technology* 32(3), **2009**, 410–417.

[74] Pröll T., Mayer K., Bolhar-Nordenkampf J., Kolbitsch P., Mattisson T., Lyngfelt A., Hofbauer H. Natural minerals as oxygen carriers for chemical looping combustion in a dual circulating fluidized bed system. *Energy Procedia* 1, **2009**, 27–34.

[75] Dueso C., Garca-Labiano F., Adnez J., de Diego L.F., Gayn P., Abad A. Syngas combustion in a chemical-looping combustion system using an impregnated Ni-based oxygen carrier. *Fuel* In Press, **2008**.

[76] Haider M., Linzer W. A One Dimensional Stationary Simulation Model for Circulating Fluidized Bed Boilers. In *Circulating Fluidized Bed Technology IV*, edited by Avidan A. AIChE, New York, **1994**, 685–692.

[77] Adanez J., Gayan P., de Diego L.F., Garcia-Labiano F., Abad A. Combustion of Wood Chips in a CFBC. Modeling and Validation. *Industrial & Engineering Chemistry Research* 42(5), **2003**, 987–999.

[78] Basu P. Combustion of coal in circulating fluidized-bed boilers: a review. *Chemical Engineering Science* 54(22), **1999**, 5547–5557.

[79] Marmo L., Rovero G., Baldi G. Modelling of catalytic gas-solid fluidised bed reactors. *Catalysis Today* 52(2-3), **1999**, 235–247.

[80] Grace J., Lim K. Reactor modeling for high-velocity fluidized beds. In *Circulating Fluidized Beds*, edited by Grace J., Avidan A., Knowlton T., chap. Reactor modeling for high-velocity fluidized beds. Chapman & Hall, New York, **1997**, first ed., 504–524.

[81] Burcat A., Ruscic B. *Third Millennium Ideal Gas and Condensed Phase Thermochemical Database for Combustion with updates from Active Thermochemical Tables*. Tech. Rep. ANL-05/20 and TAE 960, Technion-IIT, Aerospace Engineering and Argonne National Laboratory, Chemistry Division, Haifa, **1997**.

[82] Chase M. W. J., Davies C.A., Downey J. R. J., Frurip D.J., McDonald R.A., Syverud A.N. JANAF Thermochemical Tables. Third Edition. *Journal of Physical and Chemical Reference Data, Supplement* 14(1), **1985**, 1–1856.

[83] Xu J., Froment G.F. Methane steam reforming, methanation and water-gas shift: I. Intrinsic kinetics. *AIChE Journal* 35(1), **1989**, 88–96.

[84] Weimer A.W., Clough D.E. Modeling a low pressure steam-oxygen fluidized bed coal gasifying reactor. *Chemical Engineering Science* 36(3), **1981**, 549–67.

[85] Kunii D., Levenspiel O. Fluidized reactor models. 1. For bubbling beds of fine, intermediate, and large particles. 2. For the lean phase: freeboard and fast fluidization. *Industrial & Engineering Chemistry Research* 29(7), **1990**, 1226–34.

[86] Rhodes M.J., Sollaart M., Wang X.S. Flow structure in a fast fluid bed. *Powder Technology* 99(2), **1998**, 194–200.

[87] Kaushal P., Pröll T., Hofbauer H. Model development and validation: Co-combustion of residual char, gases and volatile fuels in the fast fluidized combustion chamber of a dual fluidized bed biomass gasifier. *Fuel* 86(17-18), **2007**, 2687–2695.

[88] Kaushal P., Pröll T., Hofbauer H. Model for biomass char combustion in the riser of a dual fluidized bed gasification unit: Part 1 – Model development and sensitivity analysis. *Fuel Processing Technology* 89(7), **2008**, 651–659.

[89] Kaushal P., Pröll T., Hofbauer H. Model for biomass char combustion in the riser of a dual fluidized bed gasification unit: Part II – Model validation and parameter variation. *Fuel Processing Technology* 89(7), **2008**, 660–666.

[90] Brereton C., Grace J., Yu J. Axial gas mixing in a circulating fluidized bed. In *Circulating Fluidized Bed Technology II*, edited by Basu P., Large J. Pergamon Press, Oxford, **1988**, first ed., 307–314.

[91] Patience G.S., Chaouki J. Gas phase hydrodynamics in the riser of a circulating fluidized bed. *Chemical Engineering Science* 48(18), **1993**, 3195–3205.

[92] Patience G.S., Chaouki J. Solids hydrodynamics in the fully developed region of CFB risers. In *Fluidization VIII Preprints*. Tours, France, **1995**, 33–40.

[93] Schlichthaerle P., Werther J. Axial pressure profiles and solids concentration distributions in the CFB bottom zone. *Chemical Engineering Science* 54(22), **1999**, 5485–5493.

[94] Grasa G.S., Abanades J.C., Alonso M., Gonzalez B. Reactivity of highly cycled particles of CaO in a carbonation/calcination loop. *Chemical Engineering Journal (Amsterdam, Netherlands)* 137, **2008**, 561–567.

[95] Levenspiel O. *Chemical Reaction Engineering*. John Wiley & Sons, **1999**, third ed.

Die VDM Verlagsservicegesellschaft sucht für wissenschaftliche Verlage abgeschlossene und herausragende

Dissertationen, Habilitationen, Diplomarbeiten, Master Theses, Magisterarbeiten usw.

für die kostenlose Publikation als Fachbuch.

Sie verfügen über eine Arbeit, die hohen inhaltlichen und formalen Ansprüchen genügt, und haben Interesse an einer honorarvergüteten Publikation?

Dann senden Sie bitte erste Informationen über sich und Ihre Arbeit per Email an *info@vdm-vsg.de*.

Sie erhalten kurzfristig unser Feedback!

VDM Verlagsservicegesellschaft mbH
Dudweiler Landstr. 99   Telefon +49 681 3720 174
D - 66123 Saarbrücken   Fax   +49 681 3720 1749

www.vdm-vsg.de

Die VDM Verlagsservicegesellschaft mbH vertritt

Printed by Books on Demand GmbH, Norderstedt / Germany